Guozhen Wu
Vibrational Spectroscopy

Also of Interest

Atomic Emission Spectrometry.
AES – Spark, Arc, Laser Excitation
Golloch, Joosten, Killewald, Flock, 2019
ISBN 978-3-11-052768-1, e-ISBN 978-3-11-052969-2

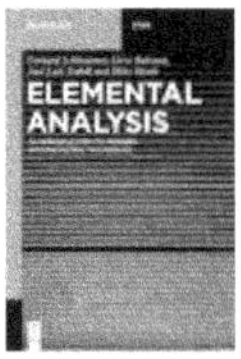

Elemental Analysis.
An Introduction to Modern Spectrometric Techniques
Schlemmer, Balcaen, Todolí, Hinds, 2018
ISBN 978-3-11-050107-0, e-ISBN 978-3-11-050108-7

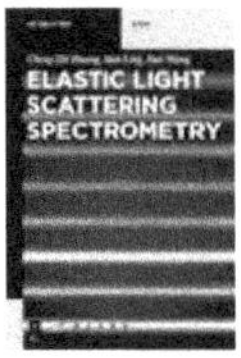

Elastic Light Scattering Spectrometry.
Huang, Ling, Wang, 2018
ISBN 978-3-11-057310-7, e-ISBN 978-3-11-057313-8

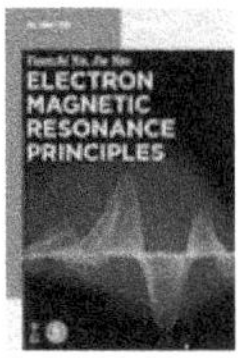

Electron Magnetic Resonance Principles.
Xu, Yao, 2019
ISBN 978-3-11-052800-8, e-ISBN 978-3-11-062010-8

Guozhen Wu

Vibrational Spectroscopy

DE GRUYTER

清華大學出版社
TSINGHUA UNIVERSITY PRESS

Author
Guozhen Wu
Department of Physics
Tsinghua University
Beijing
PR China

ISBN 978-3-11-062223-2
e-ISBN (PDF) 978-3-11-062509-7
e-ISBN (EPUB) 978-3-11-062225-6

Library of Congress Control Number: 2018967843

Bibliographic information published by the Deutsche Nationalbibliothek
The Deutsche Nationalbibliothek lists this publication in the Deutsche Nationalbibliografie; detailed bibliographic data are available on the Internet at http://dnb.dnb.de.

Typesetting: Integra Software Services Pvt. Ltd.
Cover image: Spectroscopy of Dead Sea Scrolls/Science Photo Library/Equinox Graphics
Printing and binding: CPI books GmbH, Leck

www.degruyter.com

Preface

Molecular spectroscopy is a field related to the study of the interaction of light with molecules. Its main aim is exploration of molecular structures. In low intensities, light is well described by the classical electromagnetic theory. In strong intensities, light is described by the quantum field theory. For molecules, one has to employ quantum theory to explore their structures. However, their exact solutions by quantum algorithm are usually impossible because their dynamics belong to the many-body problem. In this aspect, approximations are needed. For high excitation, sometimes, classical dynamics is useful for the study of molecular structures. We will discuss this issue in Chapter 13 of this book. Besides theoretical work, experimental work is indispensable in molecular spectroscopy. The central issue is to interpret the physical/chemical implications from the experimental data.

As mentioned, molecular dynamics is very complex. Its most distinct character is that the electron mass is much smaller than the nucleus. The consequence is that the speed of a moving electron is much faster than that of the nucleus. As molecular nuclei are moving, the electrons surrounding them can always correlate with the nuclear configuration in a molecule. On the contrary, because of excitation, electronic distribution may change suddenly with light absorption; nuclei cannot adjust their positions on time in conformity to the excited electron distribution. Because of this difference, the excitation energy of electrons is larger than that of nuclei in a molecule.

The nuclear motion can be classified as vibrational and rotational. For a linear N atomic molecule, there are 3N–5 vibrational modes, two rotational modes and three translational modes. For a nonlinear molecule, besides the 3 translational modes, there are 3N–6 vibrational modes and 3 rotational modes. The vibrational frequency is much faster than that of rotation. Therefore, we can often separate the vibrational and the rotational modes. The vibrational excitation energy is larger than that of rotation.

The contents of this book cover the topics that the author has been lecturing in the past years in the graduate school of Chinese Academy of Sciences and Tsinghua University. These topics constitute the foundation of molecular vibrational spectroscopy. As a textbook in this field, quantum theory is only introduced in Chapter 1. It would be better for the readers if they have background in quantum mechanics. However, if they lack this background then it is not fatal.

In Chapter 1, the basics of quantum mechanics are briefly introduced. Emphasis is given on the interaction of light with molecules. The physical background of a spectrum is discussed.

In Chapter 2, the separation of the electronic and nuclear motions, that is, the Born–Oppenheimer approximation is introduced. Then the potential for the nuclear motion is discussed. This potential includes the kinetic energy of electrons, the repulsive energy among the electrons, the attractive energy between electrons and

https://doi.org/10.1515/9783110625097-201

nuclei and the repulsive energy among the nuclei. With this potential energy, the vibrational and rotational motions are introduced, including their interaction with light. The spin state also affects the rotational state. This restriction on the rotational states is a natural consequence of the symmetry. The rotational and vibrational spectra are demonstrated to elucidate the molecular parameters.

In Chapter 3, the vibrational analysis is introduced. The adoption of various coordinate systems including the normal coordinate is introduced. The concept of normal mode and its physical picture is the foundation of vibrational spectroscopy.

In Chapter 4, the normal mode analysis is applied on the spectrum of thiocyanate ion adsorbed on the electrode, which is potentially dependent. This example demonstrates that fruitful information can be retrieved from the experimental spectrum by this simple but adequate method.

In Chapter 5, the application of (point) group theory on molecular spectroscopy is introduced. The group-theoretic analysis is related to the symmetry and is very powerful in the field of molecular dynamics.

In Chapter 6, the application of point group is introduced for the dynamical interpretation of solid-state spectroscopy. Besides point group, the concepts of site group and unit cell group are introduced.

In Chapter 7, the Raman effect is introduced. The essence of this effect is the coupling of the vibrational/rotational and electronic motions via light scattering. Raman effect is a two-photon process by which the electronic information can be revealed, besides the vibrational/rotational information. It can offer more information than the infra red (IR) absorption spectroscopy.

In Chapter 8, the Huckel approximation method for electronic orbitals is introduced. The group theoretic application in this field is also introduced.

In Chapter 9, the bond polarizability algorithm, proposed by the author, is introduced with its application on the spectral intensity analysis of surface-enhanced Raman effect.

In Chapter 10, the Raman excited virtual state is analyzed by the bond polarizability algorithm to reveal its electronic structure via the Raman spectral intensities. The *mystery*, if there is, of Raman excited virtual state is hence cleared.

In Chapter 11, the spectroscopic aspect of chirality is introduced. This is a new frontier. It is related to the steric structure beyond Raman and IR spectroscopies.

In Chapter 12, the bond polarizability algorithm is extended to analyze intensities of Raman optical activity. This results in a new parameter, the differential bond polarizability which characterizes the properties of chiral molecules.

In Chapter 13, the field of highly excited molecular vibration is introduced, for which the normal mode concept is no longer adequate. Instead, the nonlinear and chaotic phenomena will emerge. This approach by the classical nonlinear dynamics shows its advantage over the quantal algorithm based on the wave function. This approach is an integration of the ideas adopted from second quantization operators, Heisenberg's correspondence, Hamilton's equations of motion, pendulum dynamics, Lyapunov exponent, chaos and so on. The key experimental data are the level spacings. This classical approach is effective and offers nonlinear aspects of highly excited vibration. This chapter offers us an idea that the traditional approach based on Schrodinger equation is not an absolute necessity.

There are problem sets for some chapters. The answers are given at the end of the book. The character tables of point groups are attached in the Appendix.

The material in this book can be for a one year course. For the first semester course, any part of the 13 chapters can be chosen depending on the lecturer's will. However, Chapters 1, 2, 3, 5, 7 and 8 should be included.

The Chinese edition was published by the Tsinghua University Press in 2018. I am happy that this English edition is now being published by De Gruyter Akademie Forschung. For this, I appreciate all the efforts taken by people from both China and Germany who are involved and have made this English edition possible.

Guozhen Wu
October 15, 2018
Department of Physics, Tsinghua University

Contents

1 Foundations of quantum mechanics

In this chapter, we will first review the basic concepts of quantum mechanics and then emphasize on these topics that are the foundations for the subsequent chapters on vibrational spectroscopy.

1.1 Quantum states and operators

In quantum mechanics, a physical state can be represented by a state function $|a\rangle$ (or $\langle a|$) . $|a\rangle$ is also called a state vector. This is because that $|a\rangle$ satisfies many properties of the vectors in algebra. For example, the sum of two vectors is a vector; a vector multiplied by a number is still a vector. Correspondingly, we have the following formulae:

$$|a\rangle + |b\rangle = |c\rangle$$

$$c\,|a\rangle = |d\rangle$$

Here, $|a\rangle$, $|b\rangle$, $|c\rangle$, $|d\rangle$ are state functions, c can be any complex number.

Besides addition, there is another property: the inner product of two vectors generally denoted by "•" is a real number. Correspondingly, we can define "inner product" of two state functions, $\langle a|$ and $|b\rangle$, shown as the following:

$$\langle a|b\rangle$$

Contrarily, $\langle a|b\rangle$ can be a complex number. In the integral form, we have the following:

$$\langle a|b\rangle = \int \phi_a^* \phi_b d\tau$$

Here, ϕ_a, ϕ_b are two algebraic functions (called wave functions) for $|a\rangle$ and $|b\rangle$, respectively. The notation "*" is a complex conjugate. The integration by $d\tau$ covers all the spaces on which ϕ_a and ϕ_b are defined. From the definition, we have

$$\langle a|b\rangle = \langle b|a\rangle^*$$

Hence, in taking the inner product of two state functions, the order is important, if the product is not a real number.

The collection of all vectors together with their operators forms an algebraic structure, called the vector space. The collection of all state functions together with their operators forms an algebraic structure, called the Hilbert space.

https://doi.org/10.1515/9783110625097-001

A physical transformation is realized by an operator in quantum mechanics. A state $|a\rangle$ can be operated by an operator $\hat{T}$ and transformed to another state $|b\rangle$. That is:

$$\hat{T}|a\rangle = |b\rangle$$

If $|b\rangle$ is exactly $a|a\rangle$, then

$$\hat{T}|a\rangle = a|a\rangle$$

In such a case, a and $|a\rangle$ are called the eigenvalue and eigenfunction of $\hat{T}$, respectively. An operator can possess a set of eigenfunctions for which their inner products are all zero.

All operators T for physical transformations satisfy the following relation:

$$\langle a|\hat{T}|b\rangle = \langle b|\hat{T}|a\rangle^{\star}$$

The operators that satisfy such a property are called Hermitian operator. It can be shown that the eigenvalue of a Hermitian operator is real.

In quantum mechanics, every physical transformation is associated with an operator. Every operator possesses a simple correspondence with its classical analog, as shown in Tab. 1.1. In this table, x_k, p_{xj}, J_{xj}, H are the position, momentum, angular momentum and Hamiltonian of particle j. h is the Planck's constant, $\hbar = h/2\pi$.

Tab. 1.1: The corresponding analogs between classical and quantal operators.

Classical	Quantal
x_k	x_k
$p_{xj} = m_j v_j = m_j\left(\frac{dx_j}{dt}\right)$	$\frac{\hbar}{i}\frac{\partial}{\partial x_j}$
$J_{xj} = m_j\left(y_j\frac{dz_j}{dt} - z_j\frac{dy_j}{dt}\right)$	$\frac{\hbar}{i}\left(y_j\frac{\partial}{\partial z_j} - z_j\frac{\partial}{\partial y_j}\right)$
$H = \frac{1}{2}\sum_j \frac{1}{m_j}(p_{xj}^2 + p_{yj}^2 + p_{zj}^2) + V$	$H = -\frac{\hbar^2}{2}\sum_j \frac{1}{m_j}\nabla_j^2 + V$

The commutator between two operators $\hat{A}$ and $\hat{B}$ is defined as follows:

$$[\hat{A}, \hat{B}] = \hat{A}\hat{B} - \hat{B}\hat{A}$$

Then the following relations hold:

$$[\hat{A}, \hat{B}] = -[\hat{B}, \hat{A}]$$
$$[\hat{A}, \hat{B} + \hat{C}] = [\hat{A}, \hat{B}] + [\hat{A}, \hat{C}]$$
$$[\hat{A}, \hat{B}\hat{C}] = [\hat{A}, \hat{B}]\hat{C} + \hat{B}[\hat{A}, \hat{C}]$$

If $[\hat{A}, \hat{B}]f = 0$ for all well-defined functions f, we may simply write as follows:

$$[\hat{A}, \hat{B}] = 0$$

In this equation, $\hat{A}, \hat{B}$ are called as commute. Some operators are commute, but not all. For instance, we have the following:

$$[\boldsymbol{J}, \hat{J}^2] = 0$$

$$[\hat{J}^2, \hat{J}_+] = [\hat{J}^2, \hat{J}_-] = [\hat{J}^2, \hat{J}_z] = 0$$

$$[\hat{J}_x, \hat{J}_y] = \mathrm{i}\hat{J}_z, \quad [\hat{J}_y, \hat{J}_z] = \mathrm{i}\hat{J}_x$$

$$[\hat{J}_z, \hat{J}_-] = -\hat{J}_-, \quad [\hat{J}_+, \hat{J}_-] = 2\hat{J}_z$$

$$[\hat{x}, \hat{p}_x] = \mathrm{i}\hbar, \quad [\hat{x}, \hat{p}_x^2] = 2\hbar^2 \frac{\partial}{\partial x}$$

$$[\hat{x}, \hat{H}] = \frac{\mathrm{i}\hbar}{m}\hat{p}_x, \quad [\hat{p}_x, \hat{H}] = \frac{\hbar}{\mathrm{i}}\frac{\partial V}{\partial x}$$

$$[\hat{q}_i, G(p_1, \cdots, p_k)] = \mathrm{i}\hbar \frac{\partial G}{\partial p_i}$$

$$[\hat{p}_i, \hat{F}(q_1, \cdots, q_k)] = \frac{\hbar}{\mathrm{i}}\frac{\partial F}{\partial q_i}$$

Here,

$$\hat{\boldsymbol{J}} = \hat{J}_x\hat{\boldsymbol{x}} + \hat{J}_y\hat{\boldsymbol{y}} + \hat{J}_z\hat{\boldsymbol{z}}$$
$$\hat{J}_\pm = \hat{J}_x \pm \mathrm{i}\hat{J}_y$$

$\hat{\boldsymbol{x}}, \hat{\boldsymbol{y}}, \hat{\boldsymbol{z}}$ represent the Cartesian unit vectors.

The non-commute property is closely related to quantization and the uncertainty principle.

If two operators commute, they share a common complete set of eigenfunctions and vice versa. (Completeness means that any well-defined function can always be expressed as the linear combinations of the eigenfunctions.)

For a state $|a\rangle$, as we do a series of measurements with $\hat{A}$, we have the expectation value $\langle \hat{A} \rangle$ as follows:

$$\langle \hat{A} \rangle = \frac{\langle a|\hat{A}|a\rangle}{\langle a|a\rangle}$$

Define the mean square of deviations as $(\Delta\hat{A})^2 \equiv \left\langle \left(\hat{A} - \langle \hat{A} \rangle\right)^2 \right\rangle$, which is equal to $\langle \hat{A}^2 \rangle - \langle \hat{A} \rangle^2$.

It can be shown that for two operators, $\hat{A}, \hat{B}$, we always have

$$\Delta\hat{A}\cdot\Delta\hat{B}\geq\frac{1}{2}\left|\langle[\hat{A},\hat{B}]\rangle\right|$$

If $\hat{A},\hat{B}$ commute, then

$$\Delta\hat{A}\cdot\Delta\hat{B}\geq 0$$

Physically, this means that the measurement with these two operators can be infinitely accurate, simultaneously. Otherwise, $[\hat{A},\hat{B}]\neq 0$, $\Delta\hat{A}\cdot\Delta\hat{B}$ will be larger than a definite positive value. For instance, we have

$$\Delta\hat{J}_x\cdot\Delta\hat{J}_y\geq\frac{1}{2}\left|\langle[\hat{J}_x,\hat{J}_y]\rangle\right|=\frac{1}{2}\left|\langle\hat{J}_z\rangle\right|$$

Here, $\left|\langle\hat{J}_z\rangle\right|$ can be $l\hbar, l=0,1/2,1,3/2,2,\cdots$.

As $l\neq 0$, $\hat{J}_x,\hat{J}_y$ cannot be simultaneously measured infinitely accurately. This is the uncertainty principle.

An operator can have a differential form as shown in Tab. 1.1 It can also be expressed in a matrix form by a complete set of functions $\{|a_i\rangle\}_i$ with matrix element $T_{ij}=\langle a_i|\hat{T}|a_j\rangle$.

If $\langle a_i|a_j\rangle=\delta_{ij}$, then $\{|a_i\rangle\}_i$ satisfies the closure relation. We express it as follows:

$$\sum_i |a_i\rangle\langle a_i|=1$$

$\sum_i |a_i\rangle\langle a_i|$ is called the unit operator. It can be inserted anywhere in a series of operations. Often, this application may reduce the cumbersome manipulations in calculation (see Exercise 1.1).

If $|a_i\rangle$ is the eigenfunction of $\hat{T}$ and all the eigenfunctions are orthogonal, that is,

$$\hat{T}|a_i\rangle=a_i|a_i\rangle$$

$$T_{ij}=a_i\delta_{ij},\ \delta_{ij}=\begin{cases}0, & i\neq j\\ 1, & i=j\end{cases}$$

then matrix T is diagonal.

1.2 Time-independent perturbation

We will not go into the detailed analysis of this topic since it is treated in most of the elementary textbooks on quantum mechanics. The results are shown below.

1. Nondegenerate case

Suppose

$$H = H^0 + \lambda H', \ \lambda H' << H^0$$

H, H^0, H' are the complete Hamiltonian, Hamiltonian without perturbation and perturbation, respectively. $|n^{(0)}\rangle$, $E_n^{(0)}$, $|n\rangle$, E_n are the eigenfunctions and eigenvalues of H^0 and H. The eigenfunction and eigenvalue of H are expanded as follows:

$$|n\rangle = |n^{(0)}\rangle + \lambda|n^{(1)}\rangle + \lambda^2|n^{(2)}\rangle \cdots$$
$$E_n = E_n^{(0)} + \lambda E_n^{(1)} + \lambda^2 E_n^{(2)} + \cdots$$

with

$$|n^{(1)}\rangle = \sum_{l\neq n} \frac{|l^{(0)}\rangle\langle l^{(0)}|H'|n^{(0)}\rangle}{E_n^{(0)} - E_l^{(0)}}$$
$$E_n^{(1)} = \langle n^{(0)}|H'|n^{(0)}\rangle$$
$$E_n^{(2)} = \sum_{l\neq n} \frac{\langle n^{(0)}|H'|l^{(0)}\rangle\langle l^{(0)}|H'|n^{(0)}\rangle}{E_n^{(0)} - E_l^{(0)}}$$
$$= \sum_{l\neq n} \frac{|\langle n^{(0)}|H'|l^{(0)}\rangle|^2}{E_n^{(0)} - E_l^{(0)}}$$

2. Degenerate case

Suppose $|1^{(0)}\rangle, |2^{(0)}\rangle \ldots, |n^{(0)}\rangle$ are the degenerate states of $H^{(0)}$, the eigenvalues under the perturbation H' ($H = H^{(0)} + H'$) are the solutions of the following matrix equation:

$$\begin{vmatrix} H'_{11} - E, H'_{12}, \cdots, H'_{1n} \\ \vdots \\ H'_{n1}, \ H'_{n2}, \cdots, H'_{nn} - E \end{vmatrix} = 0$$

with

$$H'_{ij} = \langle i^{(0)}|H'|j^{(0)}\rangle$$

1.3 Time-dependent perturbation

Suppose a physical state at time t is $|n, t\rangle$, $|n, t\rangle$ satisfies the equation of motion:

$$H^0|n,t\rangle = -\frac{\hbar}{i}\frac{\partial}{\partial t}|n,t\rangle \tag{1.1}$$

If its Hamiltonian H^0 is time-independent, let

$$|n,t\rangle = \exp(-iE_n t/\hbar)|n\rangle \tag{1.2}$$

and inserting eq. (1.2) in eq. (1.1), we have

$$H^0|n\rangle = E_n|n\rangle$$

This is just the eigen-equation problem for the stationary state.

Now, under a time-dependent perturbation, $H'(t)$, the equation of motion is as follows:

$$\left(H^0 + H'(t)\right)|\phi\rangle_i = -\frac{\hbar}{i}\frac{\partial}{\partial t}|\phi\rangle_i \tag{1.3}$$

Since $\{|n,t\rangle\}_n$ is complete, $|\phi\rangle_i$ can be expressed as their linear combinations:

$$|\phi\rangle_i = \sum_n C_{ni}(t)e^{-iE_n t/\hbar}|n\rangle \tag{1.4}$$

Insert eq. (1.4) into eq. (1.3) and consider eq. (1.1), multiply both sides by $\langle m|$ and note that

$$\langle m|n\rangle = \delta_{mn}$$

then

$$\frac{d}{dt}C_{mi}(t) = -\frac{i}{\hbar}\sum_n C_{ni}(t)e^{-i(E_n - E_m)t/\hbar}\langle m|H'|n\rangle \tag{1.5}$$

$C_{mi}(t)$ shows the population of $|\phi\rangle_i$ on $|m,t\rangle$ at t.

If $\langle m|H'|n\rangle = 0$, then $C_{ni}(t)$ is constant and

$$|\phi\rangle_i = \sum_n C_{ni}|n,t\rangle$$

This means that the probability of $|\phi\rangle_i$ on $|n,t\rangle$ is time-independent. The perturbation of $H'(t)$ is ineffective on $|\phi\rangle_i$. Thus, $\langle m|H'|n\rangle$ is crucial in determining the state transition. This is the selection rule.

In the following section, we will note the effect of light by eq. (1.5).

1.4 Interaction with light

The Hamiltonian of a charged particle in a potential V under the effect of (weak) light is

$$H = \left[-\frac{\hbar^2}{2m}\nabla^2 + V(\boldsymbol{r}) \right] + \frac{ie\hbar}{mc}\boldsymbol{A}\cdot\nabla$$

Here, $\frac{ie\hbar}{mc}\boldsymbol{A}\cdot\nabla$ is because of the interaction of the particle with light, ∇ is $\frac{\partial}{\partial x}\hat{\boldsymbol{x}} + \frac{\partial}{\partial y}\hat{\boldsymbol{y}} + \frac{\partial}{\partial z}\hat{\boldsymbol{z}}$, $\boldsymbol{A}$ is the vector potential of light, its relation to the electric field $\boldsymbol{\varepsilon}$ is

$$\boldsymbol{\varepsilon}(\boldsymbol{r}, t) = -\frac{1}{c}\frac{\partial \boldsymbol{A}(\boldsymbol{r}, t)}{\partial t}$$

With the results from the last section, we have

$$H^0 = -\frac{\hbar^2}{2m}\nabla^2 + V(\boldsymbol{r})$$
$$H' = \frac{ie\hbar}{mc}\boldsymbol{A}\cdot\nabla$$

If initially, the state is $|k\rangle$, that is,

$$C_k = 1, \quad C_{l\neq k} = 0$$

From eq. (1.5), we have the following:

$$\frac{d}{dt}C_{mk}(t) = -\frac{i}{\hbar}e^{-i(E_k - E_m)t/\hbar}\left\langle m\left|\frac{ie\hbar}{mc}\boldsymbol{A}\cdot\nabla\right|k\right\rangle \tag{1.6}$$

Let

$$\boldsymbol{A} = \boldsymbol{A}_0\cos(\omega t + \boldsymbol{k}\cdot\boldsymbol{r})$$
$$= \tfrac{1}{2}\left[\boldsymbol{A}_0 e^{i(\omega t + \boldsymbol{k}\cdot\boldsymbol{r})} + c.c.\right]$$

Here, $c.c$ stands for complex conjugate. Defining $E_k - E_m = \hbar\omega_{km}$, eq. (1.6) can be written as follows:

$$\frac{d}{dt}C_{mk}(t) = a_{mk}e^{i(\omega - \omega_{km})t} + c.c. \tag{1.7}$$

with

$$a_{mk} = \frac{1}{2}\frac{e}{mc}\boldsymbol{A}_0\langle m|e^{i\boldsymbol{k}\cdot\boldsymbol{r}}\nabla|k\rangle$$

Equation (1.7) can be solved as follows:

$$C_{mk}(t) = a_{mk}\frac{e^{i(\omega - \omega_{km})t} - 1}{i(\omega - \omega_{km})} - a^*_{mk}\frac{e^{-i(\omega + \omega_{km})t} - 1}{i(\omega + \omega_{km})}$$

With the first term (with the second term, the result is the same), the probability of transition from $|k\rangle$ to $|m\rangle$ at t is

$$P_{mk}(t) = |C_{mk}(t)|^2 = 4|a_{mk}|^2 \frac{\sin^2 \frac{1}{2}(\omega - \omega_{km})t}{(\omega - \omega_{km})^2}$$

Consider the power distribution of light, $g(\omega)$

$$P_{mk}(t) = 4|a_{mk}|^2 \int_{-\infty}^{\infty} \frac{\sin^2 \frac{1}{2}(\omega - \omega_{km})t}{(\omega - \omega_{km})^2} g(\omega)d\omega \tag{1.8}$$

If the distribution is homogeneous, $g(\omega) = 1$, with

$$\int_{-\infty}^{\infty} \frac{sin^2 x}{x^2} dx = \pi$$

Equation (1.8) leads to the following:

$$P_{mk}(t) = 2\pi|a_{mk}|^2 t$$

The transition probability is linearly dependent on time.

In the following, we will explore the physical meaning of $|a_{mk}|$. For convenience, only the x direction is considered. Suppose light is along the z axis, $\boldsymbol{k} = k\hat{\boldsymbol{z}}$, then

$$\begin{aligned}(a_{mk})_x &= \frac{e}{mc} A_{0x} \left\langle m \middle| e^{ikz} \frac{\partial}{\partial x} \middle| k \right\rangle \\ &= \frac{e}{mc} A_{0x} \left\langle m \middle| (1 + ikz + \ldots) \frac{\partial}{\partial x} \middle| k \right\rangle \\ &= \frac{e}{mc} A_{0x} \left\{ \left\langle m \middle| \frac{\partial}{\partial x} \middle| k \right\rangle + \left\langle m \middle| (ikz \frac{\partial}{\partial x} \middle| k \right\rangle + \ldots \right\}\end{aligned} \tag{1.9}$$

It can be shown that (see Exercise 1.4)

$$\begin{aligned}\langle m|\tfrac{\partial}{\partial x}|k\rangle &= \tfrac{m}{\hbar^2}(E_m - E_k)\langle m|x|k\rangle \\ \langle m|z\tfrac{\partial}{\partial x}|k\rangle &= \tfrac{1}{2}\left[\langle m|z\tfrac{\partial}{\partial x} + x\tfrac{\partial}{\partial z}|k\rangle + \langle m|z\tfrac{\partial}{\partial x} - x\tfrac{\partial}{\partial z}|k\rangle\right] \\ &= \left[\tfrac{im}{k\hbar^2}(E_m - E_k)\langle m|xz|k\rangle\right] + \tfrac{i}{2\hbar}\langle m|J_y|k\rangle\end{aligned}$$

Physically, $\langle m|x|k\rangle$ is the process by the electric dipole moment, $\langle m|xz|k\rangle$ and $\langle m|J_y|k\rangle$ are the two electric quadrupole moment (or two-photon Raman) and the magnetic dipole moment, respectively.

In general, we have to consider whether the following terms that are nonzero for their corresponding processes are allowed or not.

$$\langle m|\boldsymbol{r}|k\rangle,\ \langle m|\boldsymbol{rr}|k\rangle,\ \langle m|\boldsymbol{J}|k\rangle,\ \ldots$$

We also note that

$$\langle m|\boldsymbol{r}|k\rangle \gg \langle m|\boldsymbol{rr}|k\rangle,\ \langle m|J|k\rangle,\ \ldots$$

1.5 Einstein's theory of light absorption and emission

Consider the ground and excited states $|1\rangle$, $|2\rangle$ of a quantum system as shown in Fig. 1.1. Under the effect of light with density ρ, there are three transition mechanisms: the stimulated absorption and stimulated emission between $|1\rangle$, $|2\rangle$ which are dependent on ρ. Their rates are as follows:

$$K_{12} = B_{12}\rho$$
$$K_{21} = B_{21}\rho$$

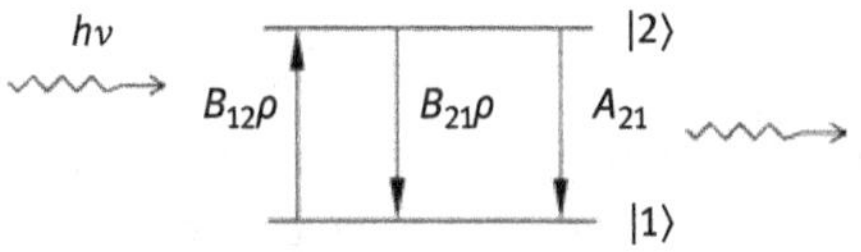

Fig. 1.1: The stimulated absorption and emission and spontaneous emission between $|1\rangle, |2\rangle$ states.

The third process is the spontaneous emission, independent of ρ. Its rate is

$$K'_{21} = A_{21}$$

Under the thermal equilibrium,

$$N_2A_{21} + N_2B_{21}\rho = N_1B_{12}\rho$$

or

$$\frac{N_2}{N_1} = \frac{B_{12}\rho}{B_{21}\rho + A_{21}} \tag{1.10}$$

N_2, N_1 are the populations of $|2\rangle$ and $|1\rangle$. By applying Boltzmann distribution law,

$$\frac{N_2}{N_1} = \frac{g_2}{g_1}e^{-h\nu/kT} \tag{1.11}$$

g_1, g_2 and $h\nu$ are the degeneracies of $|1\rangle$, $|2\rangle$ and their energy difference.

From eq. (1.10) and eq. (1.11), we derive the following:

$$\rho = \frac{A_{21}(g_2/g_1)e^{-h\nu/kT}}{B_{12} - B_{21}(g_2/g_1)e^{-h\nu/kT}} \tag{1.12}$$

According to black-body radiation,

$$\rho = \frac{8\pi h v^3}{c^3}\left(e^{hv/kT} - 1\right)^{-1} \tag{1.13}$$

By comparing eq. (1.12) and eq. (1.13), we have

$$B_{12} = B_{21}(g_2/g_1)$$
$$A_{21} = (8\pi h v^3/c^3)B_{21}$$

We note that the rates of the stimulated and spontaneous emissions are different. The light direction of the stimulated emission is the same as the original light while that of the spontaneous emission is random. Because of the spontaneous emission, the population of the excited state is very small.

Since P_{21} and $|a_{21}|^2$, P_{21} and B_{12} are proportional, B_{12} is proportional to $|a_{21}|^2$, or

$$B_{12} \sim |\langle 1|\mathbf{r}|2\rangle|^2$$

1.6 Spectral profile

The absorption spectral profile $I(\omega)$ of a medium interacting with light is the Fourier transform of the correlation function of its dipole moment μ in the time domain:

$$I(\omega) = \int_{-\infty}^{\infty} e^{-i\omega t}\langle\mu(0)\mu(t)\rangle dt \tag{1.14}$$

The average is over the medium space.

In the case of relaxation, we have

$$\mu(t) = \mu_0 e^{i\omega_0 t} e^{-rt/2} \tag{1.15}$$

Here, $\omega_0 \equiv (E_2 - E_1)/\hbar$, E_1 and E_2 are the energies of the ground and excited states. γ is the relaxation constant. From eq. (1.15) and eq. (1.14), we have

$$I(\omega) = \mu_0^2 \int_{-\infty}^{\infty} e^{-i(\omega-\omega_0)t} e^{-\gamma t/2} dt$$
$$= \mu_0^2 \frac{\gamma/2 - i(\omega-\omega_0)}{(\omega-\omega_0)^2 + \gamma^2/4}$$

The absorption coefficient K_{abs} is the real part of $I(\omega)$:

$$K_{abs} = K_0 \frac{\gamma}{(\omega-\omega_0)^2 + \gamma^2/4}$$

Its Lorentzian profile is shown in Fig. 1.2.

Often, the band width $\Delta\omega_h$ at the half maximal height which is $2K_0/\gamma$ is taken to show the profile character.

$$\Delta\omega_h = \gamma$$

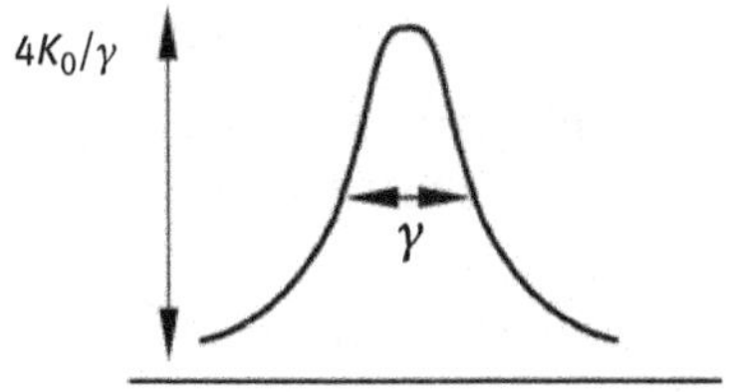

Fig. 1.2: The Lorentzian profile.

The larger the γ is, the larger the width is. $\tau = 1/\gamma$ (the dimension is time) shows the lifetime of the excited state (see Exercise 1.7). γ is related to the natural decaying constant γ_{tr} and that because of the collision γ_{coll}. Hence,

$$\gamma = \gamma_{tr} + \gamma_{coll}$$

The lifetime of the excited state τ is

$$\frac{1}{\tau} = \frac{1}{\tau_{tr}} + \frac{1}{\tau_{coll}}$$

In the room temperature, the collision rate is $10^8\ s^{-1}$ and τ_{coll} is about 10^{-8} s.

The Doppler effect which is caused because of inhomogeneity of particle velocities and is of Gaussian distribution, also affects the profile width. Its effect is

$$K_{abs}^D = K_0 \exp\left[-\frac{Mc^2}{2kT}\left(\frac{\omega - \omega_0}{\omega_D}\right)\right]^2$$

and

$$\Delta\omega_h^D = 2\omega_0 \left[2\ln 2 \frac{kT}{Mc^2}\right]^{1/2}$$

Here, M is the particle mass, c is the speed of light, T is the temperature, ω_D is the parameter related to the Doppler effect. In general, $\Delta\omega_h^D$ is two-order larger than that by the natural decaying process (see Exercise 1.7).

Because of the Doppler effect, the spectral profile is the overlap of Lorentzian curves and shows a Gaussian curve as depicted in Fig. 1.3.

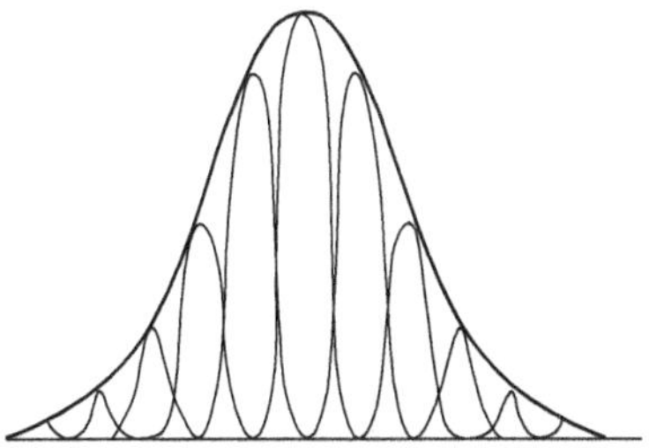

Fig. 1.3: The overlap of Lorentzian curves forms a Gaussian curve.

1.7 About wave number

Wave number (cm^{-1}) is a unit conveniently and commonly used in molecular spectroscopy to show energy. Strictly, it is not an energy unit. It is the wave numbers of the equivalent light, to which the energy is transformed, in an interval of 1 cm. Explicitly, the wave number is obtained by dividing the energy (in erg) with hc (h is the Planck's constant in erg·s and c is the speed of light in cm/s.). Since h and c are constants, the wave number is proportional to the energy. Often without misunderstanding, one calls wave number as the energy. Similar situation happens as one just calls ω as the energy or frequency, though strictly it is the angular velocity.

Exercises

1.1 As known, for the simple harmonic motion, the matrix elements of x are as follows:

$$\langle \upsilon|x|\upsilon'\rangle = 0 \quad \text{as } \upsilon' \neq \upsilon \pm 1$$

$$\langle \upsilon+1|x|\upsilon\rangle = (2\beta)^{-1/2}(\upsilon+1)^{1/2}$$

$$\langle \upsilon-1|x|\upsilon\rangle = (2\beta)^{-1/2}\upsilon^{1/2}$$

β is a constant and υ is the vibrational quantum number. Calculate

$$\langle \upsilon|x^n|\upsilon'\rangle, \mathrm{n} = 2, 3, 4, \cdots$$

1.2 In Section 1.4, if $g(\omega) = \delta(\omega - \omega_0)$, show that $P_m(t)$ is proportional to t^2 (This is the case under laser.)

1.3 Show that the transition by the electric quadrupole moment or magnetic dipole moment is of the order 10^{-6} less than that by the electric dipole moment.

1.4 Consider

$$x\varphi_k\left[\frac{d^2\varphi_m^*}{dx^2} + \frac{2m}{\hbar^2}(E_m - V(x))\varphi_m^*\right] - x\varphi_m^*\left[\frac{d^2\varphi_k}{dx^2} + \frac{2m}{\hbar^2}(E_k - V(x))\varphi_k\right] = 0$$

and integrate by parts, and show that

$$\langle m\left|\frac{\partial}{\partial x}\right|k\rangle = \frac{m}{\hbar^2}(E_m - E_k)\langle m|x|k\rangle$$

1.5 By applying Bohr's atomic model, infer that

$$\nu_{nm} \propto Z^2, |x_{nm}| \propto 1/Z \text{ and } A_{nm} \propto Z^4$$

v is the transition frequency and Z is the atomic number. Explain that the lifetime of He^+ is 1/16 of that of H.

1.6 Discuss the method to retrieve $\mu(t)$ from $I(\omega)$.

1.7 Estimate that $\Delta\omega_h^D$ is of two orders larger than that by the natural decaying process.

1.8 For the system shown in Fig. 1.4, discuss that

$$\gamma_{31} = 1/\tau_3 = A_{31} + A_{32}$$

$$\tau_{32} = 1/\tau_3 + 1/\tau_2 = A_{31} + A_{32} + A_{21}$$

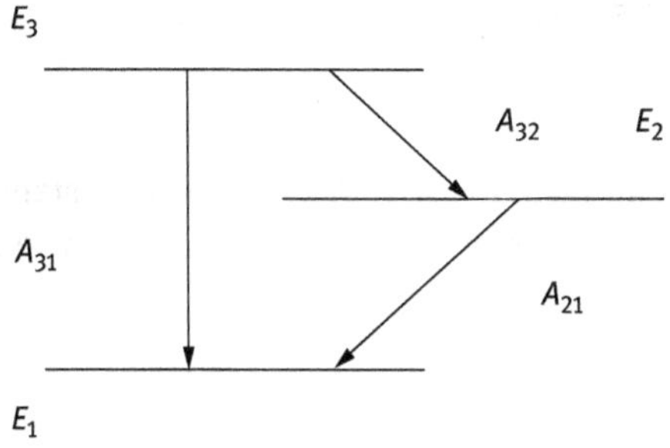

Fig. 1.4: The transitions from level 3 to level 2 and level 1.

References

[1] Steinfeld JI. Molecules and radiation. New York: Harper & Row, 1974.
[2] Levine IN. Molecular Spectroscopy. New York: John Wiley & Sons, 1975.

2 Molecular rotation

2.1 An overview

First, the frequency ratio is estimated between molecular rotation, molecular vibration and electrons circulating around the nucleus. Molecular motion as a whole is a very complicated many-body motion. However, if the frequencies of these motions are very different, we can separate them and treat them individually.

Suppose that the molecular size is a. We treat its electronic motion as the quantal one-dimensional box problem (with infinite potential at the boundary). The energy difference between the adjacent levels, E_{e1}, is

$$\hbar^2/ma^2$$

where m represents the mass of an electron and $\hbar = h/2\pi$ (h is Planck's constant).

The moment of the rotational inertia of a molecule of mass M is Ma^2. The energy difference between the adjacent levels, E_{rot}, is

$$\frac{1}{2Ma^2}[l(l+1)-(l-1)l]\hbar^2 = \frac{\hbar^2}{2Ma^2}\cdot 2l$$

$$\propto \frac{\hbar^2}{Ma^2}$$

where $l = 1, 2, 3, \ldots$.

We treat the molecular vibration as a simple harmonic oscillator with angular velocity ω. As the nucleus displaces a from its equilibrium position, the resulting potential, $\frac{1}{2}M\omega^2a^2$, has to be equal to the electronic energy difference E_{el} due to the displacement a, that is,

$$\frac{1}{2}M\omega^2a^2 = \frac{\hbar^2}{ma^2}$$

Hence, we have

$$E_{\text{vib}} = \hbar\omega \approx \frac{\hbar^2}{(mM)^{1/2}a^2}$$

From the above-mentioned results, in terms of the parameter m/M, we have

$$E_{\text{rot}} \approx (m/M)^{1/2}E_{\text{vib}} \approx (m/M)E_{\text{el}}$$

Since $m/M \approx 10^{-3}$ to 10^{-5}, the energy difference between the electronic motion, vibration and rotation is of the order of 10^2, which means that as a molecule rotates one cycle, it will vibrate 10^2 cycles and its electron will evolve the nucleus 10^4 cycles.

https://doi.org/10.1515/9783110625097-002

From this analysis, we conclude that the electronic motion can be separated from vibration and rotation. In a crude treatment, we may further separate vibration and rotation. The following section describes the electronic motion as decoupled from vibration and rotation.

2.2 Born–Oppenheimer approximation

The separation of the electronic motion from vibration and rotation is the Born–Oppenheimer approximation. The complete molecular Hamiltonian is as follows:

$$-\frac{\hbar^2}{2m}\sum_i \nabla_i^2 - \sum_A \frac{\hbar^2}{2M_A}\nabla_A^2 - \sum_{Ai}\frac{Z_A e^2}{r_{Ai}} + \sum_{A>B}\frac{Z_A Z_B e^2}{R_{AB}} + \sum_{i>j}\frac{e^2}{r_{ij}} \tag{2.1}$$

where subscripts i, A and B represent electron and nucleus; Z is the atomic number; r and R represent the interparticle distances. Suppose that the molecular wave function ψ_{mol} is the product of the electronic and nuclear wave functions, ψ_e, χ_N:

$$\psi_{\text{mol}}(r,R) = \psi_e(r,R)\,\chi_N(R) \tag{2.2}$$

From eqs. (2.1) and (2.2), the following Schrödinger equation is obtained:

$$\begin{aligned} H\psi_e\chi_N &= -\frac{\hbar^2}{2m}\sum_i \nabla_i^2\psi_e\chi_N - \sum_A \frac{\hbar^2}{2M_A}\nabla_A^2\psi_e\chi_N + \\ &\left(-\sum_{Ai}\frac{Z_A e^2}{r_{Ai}} + \sum_{A>B}\frac{Z_A Z_B e^2}{R_{AB}} + \sum_{i>j}\frac{e^2}{r_{ij}}\right)\psi_e\chi_N \\ &= E_{total}\,\psi_e\chi_N \end{aligned}$$

where E_{total} is the total energy corresponding to eq. (2.1).

From these two relations

$$\nabla_i^2\psi_e\chi_N = \chi_N\nabla_i^2\psi_e$$
$$\nabla_A^2\psi_e\chi_N = \psi_e\nabla_A^2\chi_N + 2(\nabla_A\psi_e)(\nabla_A\chi_e) + \chi_N\nabla_A^2\psi_e$$

and

$$\left\{-\frac{\hbar^2}{2m}\sum_i \nabla_i^2 - \sum_{Ai}\frac{Z_A e^2}{r_{Ai}} + \sum_{i>j}\frac{e^2}{r_{ij}}\right\}\psi_e(r,R) = E_e(R)\psi_e(r,R)$$

with $E_e(R)$ being the electronic energy when the internuclear separation is R, then we have the following:

$$\begin{gathered}\psi_e\{-\sum_A \frac{\hbar^2}{2M_A}\nabla_i^2\chi_N\}+\chi_N\{E_e\psi_e+\sum_{A>B}\frac{Z_AZ_Be^2}{R_{AB}}\psi_e\}-\\ \sum_A \frac{\hbar^2}{2m}\{2(\nabla_A\psi_e)(\nabla_A\chi_N)+\chi_N\nabla_A^2\psi_e\}\\ =E_{\text{total}}\psi_e\chi_N\end{gathered} \tag{2.3}$$

It is expected that $\nabla_A\psi_e$ and $\nabla_i\psi_e$ are of the same order. Since $-i\hbar\nabla_i$ is the momentum operator of electron, we have

$$-\frac{\hbar^2}{2M_A}\nabla_A^2\psi_e \approx \frac{P_e^2}{2M_A}=(m/M_A)\frac{P_e^2}{2m}=(m/M_A)E_e\approx 10^{-5}E_e$$

where P_e is the electron momentum. Hence, from eq. (2.3), the term

$$2(\nabla_A\psi_e)(\nabla_A\chi_N)+\chi_N\nabla_A^2\psi_e$$

can be neglected, so that eq. (2.3) reduces to

$$H_N\chi_N=\left\{\sum_A\frac{\hbar^2}{2M_A}\nabla_A^2+\sum_{A>B}\frac{Z_AZ_Be^2}{R_{AB}}+E_e(R)\right\}\chi_N=E_{\text{total}}\chi_N \tag{2.4}$$

where

$$\sum_{A>B}\frac{Z_AZ_Be^2}{R_{AB}}+E_e(R)\equiv E(R)$$

is regarded as the potential for the nuclear motion. For the diatomic molecule, the relationship between $E(R)$ and R is shown in Fig. 2.1.

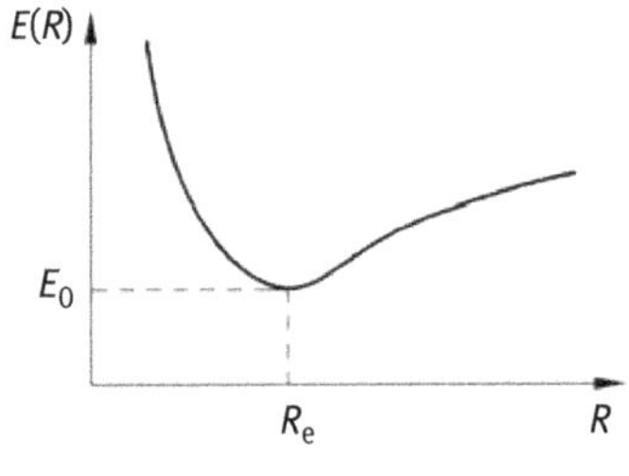

Fig. 2.1: The relationship between $E(R)$ and R for the diatomic molecule.

Expanding $E(R)$ around the equilibrium point R_e

$$E(R)=E_0+(R-R_e)\left(\frac{\partial E}{\partial R}\right)_{R_e}+\frac{1}{2}(R-R_e)^2\left(\frac{\partial^2 E}{\partial R^2}\right)_{R_e}+\cdots$$

Let $E_0 = 0$, and neglecting the higher terms, the harmonic potential is proportional to $(R-R_e)^2$:

$$E(R) \approx \frac{1}{2}\left(\frac{\partial^2 E}{\partial R^2}\right)_{Re} (R-R)^2 \equiv k\,(R-R_e)^2$$

where k is the force constant. We have to note that the force field contains the repulsion potential among nuclei, electrons, the kinetic energy of electrons and the attraction between the nuclei and electrons.

2.3 Rigid rotor

We consider the situation when molecular rotation and vibration are decoupled. Then the nuclear wave function χ_N is the product of the vibrational wave function ψ_{vib} and the rotational wave function ψ_{rot},

$$\chi_N = \psi_{\text{vib}}\psi_{\text{rot}}$$

From eq. (2.4), we have

$$-\sum_A \frac{\hbar^2}{2M_A}\nabla_A^2 \psi_{\text{vib}}\psi_{\text{rot}} = [E_{\text{total}} - E(R)]\psi_{\text{vib}}\psi_{\text{rot}} \tag{2.5}$$

For the diatomic case, the center of mass can be taken as the origin. Then $-\frac{\hbar^2}{2M_A}\nabla_A^2 - \frac{\hbar^2}{2M_B}\nabla_B^2$ reduces to $-\frac{\hbar^2}{2\mu}\nabla_R^2$ with

$$\frac{1}{\mu} = \frac{1}{M_A} + \frac{1}{M_B}, \quad \boldsymbol{R} = \boldsymbol{r}_A - \boldsymbol{r}_B$$

For the case of fixed internuclear distance at R_e, the rotational energy E_{rot} is

$$E_{\text{total}} - E(R_e) = E_{\text{rot}}$$

Then eq. (2.5) turns to be

$$-\frac{\hbar^2}{2\mu}\nabla_{R_e}^2 \psi_{\text{rot}} = E_{\text{rot}}\psi_{\text{rot}}$$

The solution is the spherical harmonics:

$$Y_J^{M_J}(\theta, \varphi) = P_J^{M_J}(\cos\theta)e^{iM_J\varphi}$$

where $P_J^{M_J}(\cos\theta)$ is the Legendre function, and $Y_J^{M_J}$ possesses the following properties:

$$\hat{J}^2 Y_J^{M_J} = \hbar^2 J(J+1) Y_J^{M_J}$$

$$\hat{J}_Z Y_J^{M_J} = M_J \hbar Y_J^{M_J}$$

where $M_J = -J, -J+1, \ldots, J$; $\hat{J}^2$ and $\hat{J}_Z$ are the operators:

$$\hat{J}^2 = -\hbar^2 \left[\left(\frac{1}{\sin\theta} \frac{\partial}{\partial\theta} \sin\theta \frac{\partial}{\partial\theta} \right) + \frac{1}{\sin^2\theta} \frac{\partial^2}{\partial\varphi^2} \right]$$

$$\hat{J}_z = \frac{\hbar}{i} \frac{\partial}{\partial\theta}$$

Rotational energy E_{rot} is

$$E_{\text{rot}} = \frac{\hat{J}^2}{2\mu R_e^2} = \frac{\hbar^2}{2\mu R_e^2} J(J+1)$$

or

$$E_{\text{rot}} = BJ(J+1)$$

where $B = \frac{\hbar^2}{2\mu R_e^2}$

We note that E_{rot} depends only on the quantum number J. The degeneracy g_J is $2J + 1$.

For a diatomic molecule, if it possesses a permanent dipole moment, μ_0, then its coupling Hamiltonian with the electric field E of light is

$$H' = -\mu_0 \bullet E = -(\mu_x E_x + \mu_y E_y + \mu_z E_z)$$

Consider the matrix element

$$\langle JM_J | H' | J'M'_J \rangle = -\mu_0 \int_0^{2\pi} \int_0^{\pi} P_J^{M_J}(\cos\theta) e^{-iM_J\phi}$$

$$\begin{pmatrix} E\sin\theta\cos\phi \\ E\sin\theta\sin\phi \\ E\cos\theta \end{pmatrix} P_J^{M_J}(\cos\theta) e^{iM_J\phi} \sin\theta d\theta d\phi \tag{2.6}$$

Equation (2.6) is nonzero in the following situations: $\mu_0 \neq 0$, $J = J' \pm 1$, $M_J = M'_J$ (electric field is along the z-direction), $M_J = M'_J \pm 1$ (electric field is along the x- and y-directions). The selection rules are as follows:

$$\Delta J = \pm 1, \Delta M_J = 0, \pm 1 \tag{2.7}$$

The physics of eq. (2.7) is that as rotation absorbs or emits one photon, its change of rotational quantum number is 1 or −1, while the change along the space axis can be 0 or ±1. The angular momentum of photon is $\pm\hbar$. Hence, the selection rules satisfy the conservation of angular momentum.

2.4 Spectral lines

Different rotational quantum numbers correspond to different rotational energy levels as shown in Fig. 2.2.

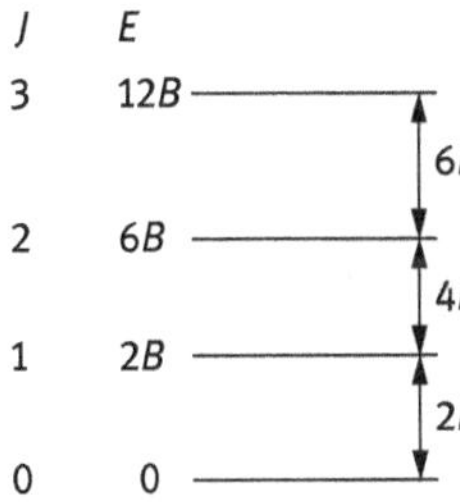

Fig. 2.2: Rotational energy levels with different J for a rigid rotor.

By the selection rule $\Delta J = \pm 1$, spectral lines are similar to the ones shown in Fig. 2.3.

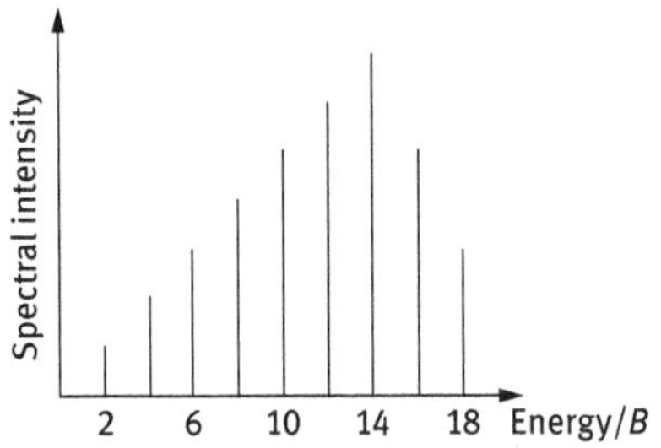

Fig. 2.3: Rotational spectral lines of a rigid rotor.

Since $\frac{B}{hc} \approx 2\,\text{cm}^{-1}$, the rotational spectrum is in the far infrared (IR) and microwave regions. Spectral line intensity depends on the state population. According to the Boltzmann distribution law, the J state population N_J is

$$N_J = (2J + 1)e^{-BJ(J+1)/kT}$$

At T = 300 K, N_J attains its maximum as J is 6. The spectral line intensities are shown in Fig. 2.3.

2.5 Symmetry

According to the Pauli exclusion principle, for a homonuclear diatomic molecule A_2, if the nuclear spin quantum number is half integer ($n/2$, n is odd), that is, A is a fermion, the sign of the molecular wave function ψ_{total} will be changed as the coordinates of its two nuclei are interchanged. Otherwise, if the nuclear spin quantum number is integer, that is, A is a boson, the sign of the molecular wave function will remain unchanged. In the following, we will show how this property affects the rotational states of a homonuclear diatomic molecule.

Consider that the nuclear spin quantum number is 1/2 and the molecular wave function ψ_{total} is the product of electronic wave function, ψ_e, vibrational wave function, ψ_{vib}, rotational wave function, ψ_{rot} and the nuclear spin wave function, $\psi_{\text{nucl. spin}}$:

$$\psi_{\text{total}} = \psi_e \psi_{\text{vib}} \psi_{\text{rot}} \psi_{\text{nucl. spin}}$$

Under the common condition, ψ_e and ψ_{vib} are in the ground states. They are symmetric under the exchange of the coordinates of the two nuclei. The situation of $\psi_{\text{nucl. spin}}$ is more complicated. It can be the following triplet state:

$$\begin{gathered} \alpha(1)\alpha(2) \\ \tfrac{1}{\sqrt{2}}[\alpha(1)\beta(2) + \alpha(2)\beta(1)] \\ \beta(1)\beta(2) \end{gathered}$$

or the singlet state

$$\frac{1}{\sqrt{2}}[\alpha(1)\beta(2) - \beta(1)\alpha(2)]$$

α, β are the wave functions of spins 1/2 and − 1/2, respectively; 1 and 2 are the labels of the two nuclei.

The triplet state is symmetric, while the singlet state is antisymmetric under the exchange of the coordinates of the two nuclei.

For the rotational wave function $Y_J^{M_J}$, the exchange of coordinates of the two nuclei is equivalent to the space inversion ($\theta \to \pi - \theta, \varphi \to \pi + \varphi$) as shown in Fig. 2.4.

Since

$$Y_J^{M_J}(\theta, \varphi) = (-1)^J Y_J^{M_J}(\pi - \theta, \pi + \varphi)$$

the rotational wave function is symmetric or antisymmetric whether J is even or odd.

For the fermion system, the molecular wave function is antisymmetric. If the nuclear spin wave function is triplet, then the rotational wave function must be

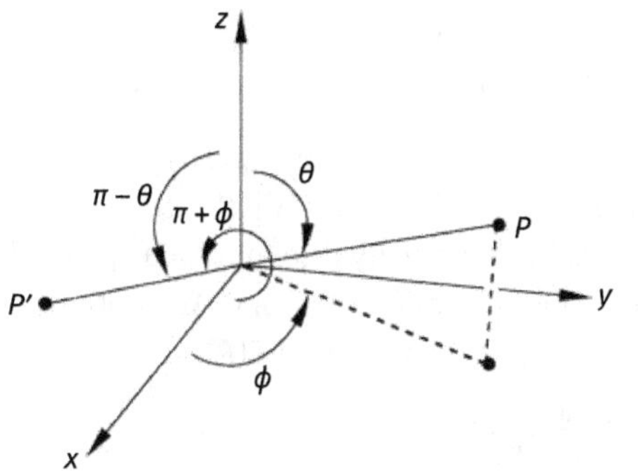

Fig. 2.4: The space inversion ($\theta \rightarrow \pi - \theta, \varphi \rightarrow \pi + \varphi$).

antisymmetric. This leads to the fact that the rotational quantum number must be odd. Those states with even J are nonexistent. On the contrary, if the nuclear spin wave function is singlet, the rotational quantum number has to be even and those states with odd J are nonexistent.

For the boson case, the analysis is similar.

This is a typical example that symmetry can impose restrictions on a physical system. In molecular spectroscopy, this situation is rather common.

2.6 Simple harmonic vibration

From Section 2.2, the wave equation of the simple harmonic vibration of a diatomic molecule is as follows:

$$\left[-\frac{\hbar^2}{2\mu}\frac{d^2}{dR^2} + E(R)\right]\psi_{\text{vib}} = E_{\text{vib}}\psi_{\text{vib}}$$

where

$$E(R) = \frac{1}{2}k\,(R - R_e)^2$$

The eigenvalues are as follows:

$$E_{\text{vib}} = \hbar\omega\left(V + \frac{1}{2}\right), V = 0, 1, 2, \ldots$$

As there is interaction with light, the perturbation Hamiltonian is as follows:

$$H' = -\mu E$$

Expanding the molecular dipole μ around $(R - R_e)$:

$$\mu = \mu_e + \left(\frac{\partial \mu}{\partial R}\right)_{Re} (R - R_e) + \frac{1}{2}\left(\frac{\partial^2 \mu}{\partial R^2}\right)_{Re} (R - R_e)^2 + \cdots$$

where μ_e is the dipole at equilibrium. If only the second-order approximation is adopted, then as $V = V' \pm 1$, $\langle V|R - R_e|V'\rangle \neq 0$ and as $V = V' \pm 2$, $\langle V|(R - R_e)^2|V'\rangle \neq 0$. The selection rules are $\Delta V = \pm 1$ or ± 2. The transition intensity due to $\Delta V = \pm 1$, called the fundamental, is larger than that due to $\Delta V = \pm 2$, called overtone. We note that overtone can also be caused by the anharmonicity of the potential $E(R)$.

Another requirement for the transition is $\left(\frac{\partial \mu}{\partial R}\right)_{Re}$ or $\left(\frac{\partial^2 \mu}{\partial R^2}\right)_{Re}$, which has to be nonzero. This means that the vibrationally induced dipole moment must be nonzero.

For the selection rule $\Delta V = \pm 1$, there is but one spectral line at

$$\Delta E = \hbar\omega$$

and it is in the IR region (200–4,000 cm^{-1}).

2.7 Vibration–rotation spectrum

When a molecule vibrates, it also rotates. For every vibrational state, it is accompanied by a series of rotational levels. Hence, a vibrational transition is accompanied by many rotational transitions. This is shown in Fig. 2.5.

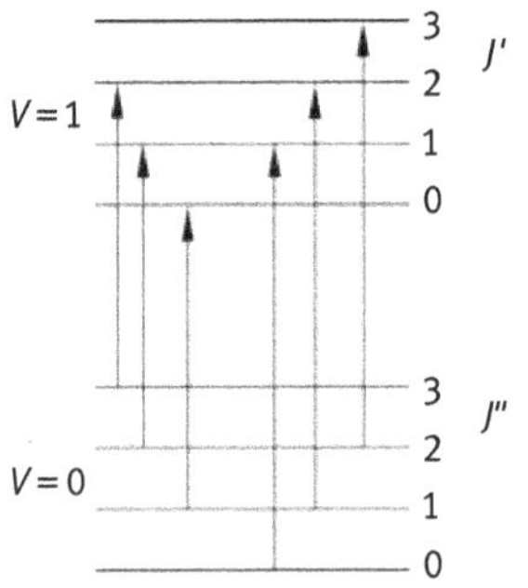

Fig. 2.5: The vibration–rotation transitions.

The rotational transitions can be $J' = J'' + 1$ or $J' = J'' - 1$. The former is called R branch and the latter is the P branch. Their frequencies are as follows:

$$\begin{aligned} \nu_R &= \nu_o + B'J'(J'+1) - B''J''(J''+1) \\ &= \nu_0 + (B' + B'')(J''+1) + (B' - B'')(J''+1)^2 \end{aligned}$$

with $J'' = 0, 1, 2, 3, \ldots$

$$\nu_P = \nu_0 + B'J'(J'+1) - B''J''(J''+1)$$
$$= \nu_0 - (B'+B'')J'' + (B'-B'')J''^2$$

with $J'' = 1, 2, 3, \ldots$.

The vibration–rotation spectrum is shown in Fig. 2.6. The position around ν_0 is called Q branch ($J'' = J'$). Often, because of symmetry, it does not show up.

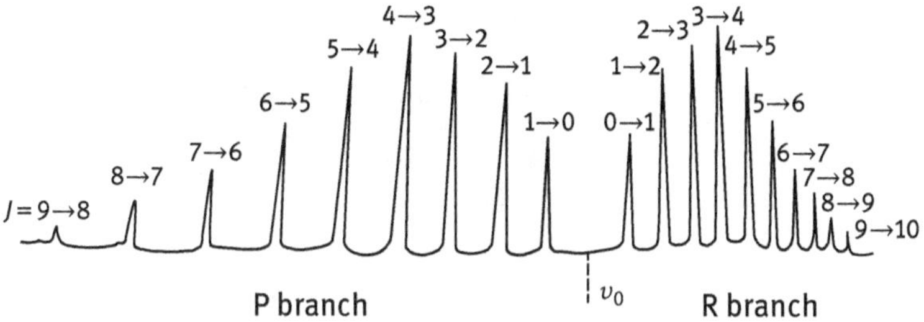

Fig. 2.6: The vibration–rotation spectrum.

Parameter B is bond length dependent. Under different vibrational quantum number V, the bond length changes, so does B. As shown earlier, at $R = R_e$,

$$B = \frac{h}{8\pi^2 c\mu R_e^2} \quad (\text{in cm}^{-1})$$

In general, we have

$$B_V = \frac{h}{8\pi^2 c\mu}\left\langle V\left|\frac{1}{R^2}\right|V\right\rangle.$$

B_V can be expanded as follows:

$$B_V = B_e - \alpha_e\left(V+\frac{1}{2}\right) + \gamma_e\left(V+\frac{1}{2}\right)^2 + \delta_e\left(V+\frac{1}{2}\right)^3 + \cdots$$

It is shown that [1]

$$-\frac{\alpha_e}{B_e} = \frac{6B_e}{\omega_e}$$

Since $B_e \approx 1\ \text{cm}^{-1}, \omega_e \approx 10^3\ \text{cm}^{-1}$, $-(\alpha_e/B_e) \approx -0.6\%$.

This means that the change in B due to vibration is about 0.6%. For larger V, R is larger and B is smaller.

2.8 Centrifugal effect

When a molecule rotates, the caused centrifugal force F_c is

$$F_c = \frac{\hat{J}^2}{\mu R^3}$$

For quantum number J, F_c is

$$\frac{J(J+1)\hbar^2}{\mu R^3}$$

There is the balance between the centrifugal force and the bond attraction force F_r. By Hooke's law:

$$F_r = k(R - R_e)$$

Thus, we have

$$R - R_e = \frac{J(J+1)\hbar^2}{\mu R^3 k}$$

Then the overall rotational energy has to include the term $\frac{1}{2}k(R-R_e)^2$

$$E_{\text{rot}} = \frac{J(J+1)\hbar^2}{2\mu R^2} + \frac{1}{2}k(R-R_e)^2$$

$$= \frac{J(J+1)\hbar^2}{2\mu R^2} + \frac{J^2(J+1)^2\hbar^4}{2\mu^2 R^6 k}$$

or

$$BJ(J+1) + D[J(J+1)]^2$$

$D[J(J+1)]^2$ is due to the centrifugal effect.

2.9 Anharmonicity

Harmonic model for vibration is a crude approximation. At high excitation, anharmonic effect will be evident. We may express vibrational energy E_V in terms of $V + \frac{1}{2}$ as follows:

$$E_V = \omega_e\left(V+\frac{1}{2}\right) - \omega_e\chi_e\left(V+\frac{1}{2}\right)^2 + \omega_e y_e\left(V+\frac{1}{2}\right)^3 + \omega_e z_e\left(V+\frac{1}{2}\right)^4$$

The energy difference is

$$\Delta E_V = E_{V+1} - E_V$$
$$= \omega_e - 2(V+1)\omega_e \chi_e + \cdots$$

As the bond is close to dissociation $\Delta E_V = 0$. Hence, we obtain

$$V_{\max} = \frac{1}{2x_e} - 1$$

Often, $V_{\max}$ is obtained by extrapolation as shown in Fig. 2.7. In general, $V_{\max}$ is between 60 and 100.

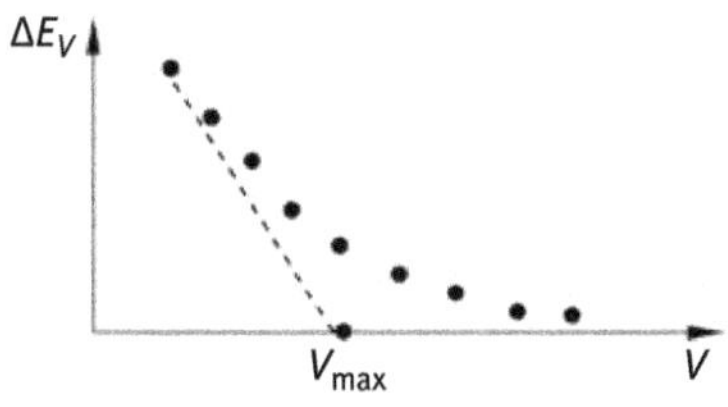

Fig. 2.7: $V_{\max}$ is obtained by extrapolation.

The bond dissociation energy is as follows:

$$D_e = \sum_{V=0}^{V_{\max}} \Delta E_v$$

2.10 Rotational spectrum of polyatomic molecules

It can be shown that for polyatomic molecules, principal axes can be found, where the rotational energy is written as follows:

$$E_{\text{rot}} = \tfrac{1}{2}\left(I_a\omega_a^2 + I_b\omega_b^2 + I_c\omega_c^2\right)$$
$$= \frac{J_a^2}{2I_a} + \frac{J_b^2}{2I_b} + \frac{J_c^2}{2I_c}$$

where I_a, I_b, I_c are the rotational inertia moments along the principal axes. Since $J^2 = J_a^2 + J_b^2 + J_c^2$, we obtain the following:

$$E_{\text{rot}} = \frac{1}{2}\left(\frac{J^2}{I_b} - \frac{J_a^2}{I_b} - \frac{J_c^2}{I_b} + \frac{J_a^2}{I_a} + \frac{J_c^2}{I_c}\right)$$

where J_a, J_b, J_c, J are the rotational quantum numbers along the principal axes a, b, c and the total rotational quantum number.

Often we take $I_a < I_b < I_c$ and classify the cases as follows:

1. Prolate top $I_a < I_b = I_c$
2. Oblate top $I_a = I_b < I_c$

3. Spherical top $I_a = I_b = I_c$
4. Asymmetric top $I_a < I_b < I_c$

For the prolate top

$$E_{\text{rot}} = \frac{1}{2}\left(\frac{J^2}{I_b} - \frac{J_a^2}{I_b} + \frac{J_a^2}{I_a}\right)$$

By quantization,

$$J^2 \to \hat{J}^2$$
$$J_a^2 \to \hat{J}_a^2$$

Eigenvalues of $\hat{J}^2, \hat{J}_a^2$ are $J(J+1)\hbar^2$ and $K^2\hbar^2$ with $K = -J, -J+1, \ldots, J$. Then

$$E_{\text{rot}} = \frac{J(J+1)\hbar^2}{2I_b} + \frac{K^2\hbar^2}{2}\left(\frac{1}{I_a} - \frac{1}{I_b}\right)$$

Similarly, for the oblate top, its rotational energy is as follows:

$$E_{\text{rot}} = \frac{J(J+1)\hbar^2}{2I_b} + \frac{K^2\hbar^2}{2}\left(\frac{1}{I_c} - \frac{1}{I_b}\right)$$

For the asymmetric top, the situation is more complicated. For more details, see ref. [2].

The rotational wave functions for the prolate and oblate tops are of the form

$$\psi_{JKM} \propto P_J^{M_J}(\cos\theta)e^{iK\chi}e^{iM\varphi}$$

where θ and φ are the spherical coordinates of the experimental laboratory, χ is the projection angle for I_a (or I_c). Since K and M share the same mathematical form, their selection rules are the same:

$$\Delta K = 0, \ \pm 1$$

Pictorially, we may visualize M as the projected quantum number along the z-axis of the experimental laboratory and K the projection along the molecular short (or long) principal axis.

Exercises

2.1 Suppose that the vibrational potential of a diatomic molecule is $E(R) = D_e\left[1 - e^{-\beta(R-R_e)}\right]^2$ and that the average kinetic energy E_k and the average potential E_p at R is

$$E_k = -E - R\frac{dE}{dR}$$
$$E_p = 2E + R\frac{dE}{dR}$$

(1) Plot $E(R)$, E_k and E_p as functions of R
(2) As $\frac{d^2E}{dR^2} = 0$, show that $R = R_e + \frac{1}{\beta}\ln 2$
(3) As $R_e < R < R_e + \frac{1}{\beta}\ln 2$ and $R > R_e + \frac{1}{\beta}\ln 2$, show the behavior of E_k and E_p.

2.2 At $T \approx 300$ K, in Section 2.4, show that N_J attains its maximum as $J \approx 6$.

2.3 Not for all the diatomic molecules, their wave functions of the electronic ground state are symmetric under the permutation of their two nuclear coordinates, for example, oxygen molecule. The nuclear spin quantum numbers of ^{16}O and ^{17}O are 0 and 5/2, respectively. Infer the allowed J in $^{16}O_2$ and $^{17}O_2$.

2.3 In P and R branches, the spectral line spaces vary. Show how the line space becomes larger in the P branch, as J'' increases. While in the R branch, the situation is opposite.

2.4 The vibrational transitions ($0 \rightarrow V$) of H_2 are shown in Tab. 2.1.

Tab. 2.1: The vibrational transitions ($0 \rightarrow V$) of H_2.

V	ΔE (cm^{-1})	V	ΔE (cm^{-1})
1	4,161.14	8	26,830.97
2	8,087.11	9	29,123.93
3	11,782.35	10	31,150.19
4	15,250.36	11	32,886.85
5	18,491.92	12	34,301.83
6	20,505.65	13	35,351.01
7	24,287.83	14	35,972.97

Calculate x_e, ω_e, V_{max} [3, 4].

References

[1] Herzberg G. Molecular Spectroscopy and Molecular Structure: Spectra of Diatomic Molecules. New York: D. Van Nostrand, 1953, 1: 108.

[2] Allen HC, Cross PC. Molecular Vib-rotors: The theory and interpretation of high resolution infrared spectra. New York: Wiley, 1963.

[3] Beuter H. J. Phys. Chem. 1934, 27B: 287.

[4] Herzberg G, Howe LL. Can. J. Phys. 1959, 37:636.

3 Molecular vibrations

3.1 Normal vibrational modes

For a molecule possessing N atoms, the total degrees of freedom is $3N$, where three degrees are for the translation and the other three (for nonlinear molecules) or two (for linear molecules) degrees are for the rotation. Hence, the vibrational degrees of freedom are $3N$–6 or $3N$–5.

The kinetic energy can be written as follows:

$$T = \frac{1}{2}\sum_{\alpha-1}^{N} m_\alpha \left[\left(\frac{d\Delta x_\alpha}{dt}\right)^2 + \left(\frac{d\Delta y_\alpha}{dt}\right) + \left(\frac{d\Delta z_\alpha}{dt}\right)^2\right] \tag{3.1}$$

where $\Delta x_\alpha, \Delta y_\alpha, \Delta z_\alpha$ are defined as follows:

$$\Delta x_\alpha = x_\alpha - (x_\alpha)_e$$

$$\Delta y_\alpha = y_\alpha - (y_\alpha)_e$$

$$\Delta z_\alpha = z_\alpha - (z_\alpha)_e$$

$(x_\alpha)_e, (y_\alpha)_e, (z_\alpha)_e$ are the coordinates of the atom α at equilibrium and $x_\alpha, y_\alpha, z_\alpha$ are its instantaneous coordinates. m_α is its mass.

For convenience, we define a new set of coordinates q_i:

$$q_1 = \sqrt{m_1}\Delta x_1, \quad q_2 = \sqrt{m_1}\Delta y_1, \quad q_3 = \sqrt{m_1}\Delta z_1, \quad q_4 = \sqrt{m_2}\Delta x_2, \quad \ldots$$

then eq. (3.1) becomes

$$T = \frac{1}{2}\sum_{i}^{3N} q_i^2 \tag{3.2}$$

The potential energy V is a function of q_i:

$$V = V(q_1, \cdots, q_{3N})$$

The expansion in terms of q_i leads to

$$V = V_0(0, \cdots, 0) + \sum_{i=1}^{3N}\left(\frac{\partial V}{\partial q_i}\right)_0 q_i + \frac{1}{2}\sum_{i=1}^{3N}\sum_{j=1}^{3N}\left(\frac{\partial^2 V}{\partial q_i \partial q_j}\right)_0 q_i q_j + \cdots \tag{3.3}$$

At equilibrium, the potential is the least, then

$$\left(\frac{\partial V}{\partial q_i}\right)_0 = 0$$

https://doi.org/10.1515/9783110625097-003

Meanwhile, we define

$$V_0 = 0$$

and keep only the second-order terms (the harmonic approximation), and then eq. (3.3) becomes

$$V = \frac{1}{2}\sum_i \sum_j f_{ij} q_i q_j \tag{3.4}$$

where

$$f_{ij} \equiv \left(\frac{\partial^2 V}{\partial q_i \partial q_j}\right)_0$$

is the force constant.

The kinetic energy T and potential energy V are now the functions of $\dot{q}$ and q. Equations of motion for the Lagrangian $L = T - V$ are as follows:

$$\frac{d}{dt}\left(\frac{\partial L}{\partial \dot{q}_i}\right) - \frac{\partial L}{\partial q_i} = 0, \; i = 1, 2, \cdots, 3N$$

or

$$\ddot{q}_i + \sum_{j=1}^{3N} f_{ij} q_j = 0, \quad \mathrm{i} = 1, 2, \cdots, 3N \tag{3.5}$$

Now suppose that

$$q_i = q_i^0 \cos(\omega t + \varepsilon)$$

and substituting it into eq. (3.5), we have

$$\sum_{j=1}^{3N} (f_{ij} - \delta_{ij}\omega^2) q_j^0 = 0, i = 1, 2, \cdots 3N \tag{3.6}$$

For the equation set to have nonzero solutions, it is required that

$$\begin{vmatrix} f_{11} - \omega^2, & f_{12}, & \ldots, & f_{1,3N} \\ f_{21}, & f_{22} - \omega^2, & \ldots, & f_{2,3N} \\ \vdots & & & \\ f_{3N,1} & \ldots, & \ldots, & f_{3N,3N} - \omega^2 \end{vmatrix} = 0$$

or in simple notation:

$$\det\left|f_{ij} - \delta_{ij}\omega^2\right| = 0 \tag{3.7}$$

This determinant has $3N$ solutions. Suppose that its solutions are ω_k^2 s, then the relative $q_{i,k}^0$ can be obtained. By the normalization condition,

$$\sum_i {q_{i,k}^0}^2 = 1 \tag{3.8}$$

$q_{i,k}^0$ can be determined. The normalized $q_{i,k}^0$ is expressed as L_{ik}.

To each ω_k, there is a corresponding set of L_{ik}. The motion is such that all the atoms possess the same frequency, ω_k, with various amplitudes, which are L_{ik}s. Such a motion is called the normal mode. Equation (3.7) has 6 (or 5) zero solutions corresponding to the translational and rotational motions.

For $\omega_k \neq 0$, the general solution of eq. (3.5) is

$$q_i = \sum_{k=1}^{3N-6} C_k L_{ik} \cos(\omega_k t + \varepsilon) \tag{3.9}$$

with

$$i = 1, 2, \cdots, 3N-6 \quad (\text{or } 3N-5)$$

C_ks are the arbitrary constants.

3.2 Normal coordinates

Besides q_{ik}, normal coordinate Q_k is defined, by which the kinetic and potential energies are given as follows:

$$T = \frac{1}{2}\sum_{k=1}^{3N-6} \dot{Q}_k^2, \; V = \frac{1}{2}\sum_{k=1}^{3N-6} \lambda_k^2 Q_k^2 \tag{3.10}$$

or in matrix notation

$$T = \tfrac{1}{2}\boldsymbol{P}_Q^{\mathrm{T}}\boldsymbol{P}_Q$$
$$V = \tfrac{1}{2}\boldsymbol{Q}^{\mathrm{T}}\boldsymbol{\Lambda}\boldsymbol{Q}$$

where

$$\Lambda_{ki} = \lambda_k^2 \delta_{ki}$$

$$\boldsymbol{P}_Q = \begin{bmatrix} \dot{Q}_1 \\ \dot{Q}_2 \\ \vdots \\ \dot{Q}_{3N-6} \end{bmatrix}$$

$$\boldsymbol{P}_Q^{\mathrm{T}} = [\dot{Q}_1, \dot{Q}_2, \cdots \dot{Q}_{3N-6}]$$

$$\boldsymbol{Q} = \begin{bmatrix} Q_1 \\ Q_2 \\ \vdots \\ Q_{3N-6} \end{bmatrix}$$

$$\boldsymbol{Q}^{\mathrm{T}} = [Q_1, Q_2, \cdots Q_{3N-6}]$$

There is a linear transformation between q_i and Q_k:

$$q_i = \sum_{k=1}^{3N-6} l_{ik} Q_k \tag{3.11}$$

By eq. (3.10) and the Lagrangian equation, the equation for Q_k is

$$\ddot{Q}_k + \lambda_k^2 Q_k = 0 \tag{3.12}$$

Suppose that its solution is

$$Q_k = Q_k^0 \cos(\lambda_k t + \varepsilon)$$

By eq. (3.11), we have

$$q_i = \sum_{k}^{3N-6} l_{ik} Q_k^0 \cos(\lambda_k t + \varepsilon)$$

By comparing with eq. (3.9), we can have $l_{ik} = L_{ik}, \lambda_k = \omega_k$. This shows that the motion with Q_k as the coordinate has frequency, ω_k. Q_k and q_i are then related by the following equation:

$$q_i = \sum_{k=1}^{3N-6} L_{ik} Q_k$$

or in matrix notation:

$$\boldsymbol{q} = \boldsymbol{L}\boldsymbol{Q}$$

$$\boldsymbol{Q} = \boldsymbol{L}^{-1}\boldsymbol{q}$$

From eq. (3.10), Schrodinger's equation for the molecular vibration is given as follows:

$$\sum_{k=1}^{3N-6} \left[-\frac{\hbar^2}{2}\frac{\partial^2}{\partial Q_k^2} + \frac{1}{2}\omega_k^2 Q_k^2 \right] \psi_{\text{vib}} = E_{\text{vib}} \psi_{\text{vib}} \tag{3.13}$$

By the separation of variables, we have

$$\psi_{\text{vib}} = \phi_{V_1}(Q_1)\phi_{V_2}(Q_2)\ldots\phi_{V_{3N-6}}(Q_{3N-6}) \equiv |V_1\, V_2 \ldots V_{3N-6}\rangle$$

Then, from eq. (3.13), we have the following:

$$\left(-\frac{\hbar^2}{2}\frac{\partial^2}{\partial Q_k^2} + \frac{1}{2}\omega_k^2 Q_k^2\right)\phi_{V_k}(Q_k) = E_{V_k}\phi_{V_k}(Q_k)$$

with

$$k = 1, 2, \cdots, 3N-6$$

The eigenenergy and eigenfunction of the harmonic motion are given as follows:

$$E_{v_k} = \hbar\omega_k\left(V_k + \frac{1}{2}\right)$$

$$\varphi_{v_k}(Q_k) = N_{v_k} e^{-\alpha_k Q_k^2/2} H_{v_k}(\sqrt{\alpha_k} Q_k)$$

where $\alpha_k = \omega_k/\hbar^2$, $H_{v_k}(\sqrt{\alpha_k}Q_k)$ is the Hermite polynomial and V_k is the quantum number. Hence, the vibrational energy, E_{vib} is given as follows:

$$\sum_k E_{v_k} = \sum_k \hbar\omega_k\left(V_k + \frac{1}{2}\right)$$

In summary, we have the following important concepts about molecular vibration: Under the harmonic approximation, molecular vibration is decomposed as a set of normal modes where all the atoms vibrate around their equilibrium positions in phase or out of phase π with the same frequency, though with different amplitudes. All the normal modes are independent. When they are expressed in the normal coordinates, they are simply one-dimensional simple harmonic oscillator.

Each set of $(V_1, \ldots, V_{3N-6})$ corresponds to a level. The transitions can be of different combinations. For example, for a triatomic molecule, there are three modes. The transitions can be

(000) → (100) fundamental transition
(100) → (010) difference transition
(100) → (200) hot transition
(000) → (200) overtone transition
(000) → (110) combination transition

Certainly, there can be many other combinations.

Note we have (see Exercise 3.1, under the specific condition $L^{-1} = L^{\mathrm{T}}$)

$$Q_k = \sum_i L_{ik} q_i$$

Hence, if arrows are used to depict the directions and magnitudes (L_{ik}) of the movement of atoms in a normal mode, the same figure can be used to show the transformation between Q_k and q_i. For example, H_2O has three normal modes with frequencies: $\omega_1 = 3657 cm^{-1}, \omega_2 = 1595 cm^{-1}, \omega_3 = 3756 cm^{-1}$. They can be depicted as shown in Fig. 3.1. Of course, L_{ik} has to be known for depicting Fig. 3.1. The method for calculating L_{ik}, is called the *normal mode analysis.*

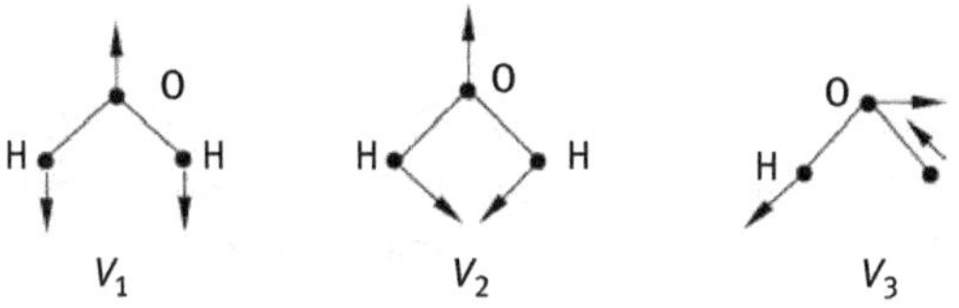

Fig. 3.1: The three normal modes of H_2O.

3.3 Selection rules

The selection rule for the transition from the initial state

$$|\psi_I\rangle = |V_1\, V_2 \cdots V_{3N-6}\rangle$$

to the final state

$$|\psi_F\rangle = |V'_1\, V'_2 \cdots V'_{3N-6}\rangle$$

depends on whether the transition moment integral

$$\mu_{FI} = \langle\psi_F|\mu|\psi_I\rangle$$

is zero or not. Here, μ is the electric moment. $\mu = \mu(Q)$ and we may have

$$\mu(Q) = \mu(0) + \sum_{k=1}^{3N-6} (\partial\mu/\partial Q_k)_0 Q_k + \cdots$$

Neglecting the higher terms and supposing that $\mu(0)=0$, then

$$\mu(Q) = \sum_{k=1}^{3N-6} (\partial\mu/\partial Q_k)_0 Q_k$$

$$\mu_{FI} = \sum_{k=1}^{3N-6} (\partial\mu/\partial Q_k)_0 < V'_k|Q_k|V_k > \cdot \prod_{j\neq k} < V'_j|V_j >$$

For $\mu_{FI} \neq 0$, for the specific mode m, we require that

$$(\partial\mu/\partial Q_m)_0 \neq 0$$

and $\Delta V_m = \pm 1$, as $j \neq m$, $\Delta V_j = 0$.

This means that in the harmonic approximation, when a molecule interacts with light, only one of its modes can absorb/emit a photon and the change of the quantum number is 1.

3.4 Generalized coordinates

In Section 3.1, we have defined two coordinates: q_i and Q_k. q_i is intuitive, but the analysis based on it is very cumbersome. Q_k is more convenient, though its configuration can only be recognized after the normal mode analysis. In practice, the realization of Q_k from q_i is not often adopted; instead, other coordinates are considered. For this, the concept of generalized coordinates is required.

Suppose that the generalized coordinate is S_t (which can be q_i or Q_k).

Let the coordinate ξ_i be

$$\xi_1 = \Delta x_1, \quad \xi_2 = \Delta y_1, \quad \xi_3 = \Delta z_1, \quad \xi_4 = \Delta x_2, \quad \cdots$$

The transformation from ξ_i to S_t is

$$S_t = \sum_{i-1}^{3N-6} B_{ti}\xi_i$$

The kinetic energy T is

$$T = \frac{1}{2}\sum_{i-1}^{3N-6} m_i \dot{\xi}_i^2$$

The momentum P_i corresponding to ξ_i is

$$P_i = \frac{\partial T}{\partial \dot{\xi}_i} = \sum_i \frac{\partial T}{\partial \dot{S}_t}\frac{\partial \dot{S}_t}{\partial \dot{\xi}_i} = \sum_t P_t B_{ti}$$

where P_t is the momentum corresponding to $\dot{S}_t$.

The kinetic energy can be written as follows:

$$T = \frac{1}{2}\sum_i \frac{P_i^2}{m_i} = \frac{1}{2}\sum_i \frac{1}{m_i}\left(\sum_t P_t B_{ti}\right)^2$$

$$= \frac{1}{2}\sum_{tt'}\left(\sum_i \frac{1}{m_i} B_{ti}B_{t'i}\right)P_t P_{t'}$$

Defining

$$G_{tt'} = \sum_i \frac{1}{m_i} B_{ti}B_{t'i}$$

then

$$T = \frac{1}{2}\sum_{tt'} G_{tt'}P_tP_{t'}$$

Similarly, potential V can be expanded in S_t as follows:

$$V = \frac{1}{2}\sum_{tt'} F_{tt'}S_tS_{t'}$$

In matrix notation, we have the following:

$$T = \tfrac{1}{2}\boldsymbol{P}^{\mathrm{T}}\boldsymbol{GP}$$
$$V = \tfrac{1}{2}\boldsymbol{S}^{\mathrm{T}}\boldsymbol{FS}$$

From Section 3.2, in terms of Q and $\boldsymbol{P}_Q(=\dot{\boldsymbol{Q}})$

$$T = \tfrac{1}{2}\boldsymbol{P}_Q^{\mathrm{T}}\boldsymbol{P}_Q$$
$$V = \tfrac{1}{2}\boldsymbol{Q}^{\mathrm{T}}\boldsymbol{\Lambda Q}$$

If the transformation from $\boldsymbol{Q}$ to $\boldsymbol{S}$ is

$$\boldsymbol{S} = \boldsymbol{LQ}$$

then

$$V = \frac{1}{2}\boldsymbol{S}^{\mathrm{T}}\boldsymbol{FS} = \frac{1}{2}\boldsymbol{Q}^{\mathrm{T}}\boldsymbol{L}^{\mathrm{T}}\boldsymbol{FLQ} = \frac{1}{2}\boldsymbol{Q}^{\mathrm{T}}\boldsymbol{\Lambda Q}$$

Hence,

$$\boldsymbol{\Lambda} = \boldsymbol{L}^{\mathrm{T}}\boldsymbol{FL}$$

On the other hand, we have

$$P_{Q_k} = \frac{\partial T}{\partial \dot{Q}_k} = \sum_t \frac{\partial T}{\partial \dot{S}_t}\frac{\partial \dot{S}_t}{\partial \dot{Q}_k} = \sum_t P_tL_{tk}$$

or in the matrix form

$$\boldsymbol{P}_Q = \boldsymbol{L}^{\mathrm{T}}\boldsymbol{P}$$

Combining the expression

$$T = \frac{1}{2}\boldsymbol{P}_Q^{\mathrm{T}}\boldsymbol{P}_Q$$

we have

$$\boldsymbol{T} = \frac{1}{2}\boldsymbol{P}^{\mathrm{T}}\boldsymbol{LL}^{\mathrm{T}}\boldsymbol{P}$$

Since

$$T = \frac{1}{2}\boldsymbol{P}^{\mathrm{T}}\boldsymbol{G}\boldsymbol{P}$$

then $\boldsymbol{G} = \boldsymbol{L}\boldsymbol{L}^{\mathrm{T}}$

or

$$\boldsymbol{L}^{\mathrm{T}} = \boldsymbol{L}^{-1}\boldsymbol{G}$$

With the expression,

$$\boldsymbol{\Lambda} = \boldsymbol{L}^{\mathrm{T}}\boldsymbol{F}\boldsymbol{L}$$

we have

$$\boldsymbol{\Lambda} = \boldsymbol{L}^{-1}\boldsymbol{G}\boldsymbol{F}\boldsymbol{L}$$

Obviously, $\boldsymbol{GF}$ is diagonalized by $\boldsymbol{L}$ through a similarity transformation.

Rewriting the expression, we have

$$\boldsymbol{G}\boldsymbol{F}\boldsymbol{L} = \boldsymbol{L}\boldsymbol{\Lambda}$$

or

$$\sum_{t'} (GF)_{tt'} L_{t'k} = L_{tk}\Lambda_{kk}$$

or

$$\sum_{t'} \left[(GF)_{tt'} - \delta_{tt'}\omega_k^2\right] L_{t'k} = 0$$

The normal mode analysis [1] is first to construct $\boldsymbol{G}$ by its definition, then to obtain $\boldsymbol{F}$ matrix and finally to diagonalize $\boldsymbol{GF}$. The diagonal elements are ω_k^2 and the eigenvectors are $\boldsymbol{L}$ from which $\boldsymbol{Q}$ can be obtained (note that $\boldsymbol{L}^{-1}\boldsymbol{S}=\boldsymbol{Q}$).

Bond angle and length can be chosen for S_t. This is the internal coordinate and is generally denoted by R.

For example, for H_2O there can be two sets of internal coordinates as shown in Fig. 3.2. Note that α, r_1, r_2 and r_3 are referred as the variation of bond angle and

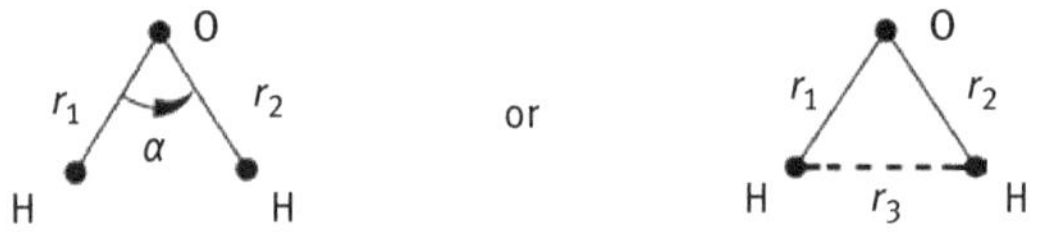

Fig. 3.2: The internal coordinates for H_2O.

length, respectively. In terms of $\{r_1, r_2, \alpha\}$, the potential energy V can be written as follows:

$$V = \frac{1}{2}\left(F_{11}r_1^2 + F_{22}r_2^2 + F_{33}\alpha^2 + 2F_{12}r_1r_2 + 2F_{13}r_1\alpha + 2F_{23}r_2\alpha\right)$$

with

$$F_{tt'} = \left(\frac{\partial^2 V}{\partial S_t \partial S_{t'}}\right)_0$$

If the cross terms, F_{12}, F_{13}, F_{23} are neglected, and

$$F_{11} = F_{22} = F_r$$

is noted, then we have the following:

$$V = \frac{1}{2}\left(F_r r_1^2 + F_r r_2^2 + F_\alpha \alpha^2\right)$$

This is the valence bond force field approximation.

For $\boldsymbol{G}$, it can be written by its definition:

$$\boldsymbol{G}_{tt'} = \sum_i \frac{1}{m_i} B_{ti} B_{t'i}$$

This algorithm is very cumbersome. Other techniques can be more convenient. This is discussed in Sections 4.1 and 4.2 and Appendix VI of Ref. 3.1.

Exercise 3.1 shows the relationship between the various coordinate systems.

For a molecule of N atoms, bond stretches and bendings can be adopted as the internal coordinates $\boldsymbol{R}$, where there are often the redundant coordinates. That is, the number of the internal coordinates chosen is larger than the number of the normal modes, $3N$–6. To delete the redundant coordinates, one may adopt a set of $3N$–6 independent and irredundant coordinates according to the structure and symmetry of the molecule. These coordinates are often the symmetry coordinates (see Section 5.14).

Suppose that the number of the internal coordinates $\boldsymbol{R}$ is m, which is larger than $3N$–6. Their relation with the symmetry coordinates $\boldsymbol{S}$ is $\boldsymbol{S} = \boldsymbol{B}_S\boldsymbol{R}$. The transformation between the force fields $\boldsymbol{F}_R$ and F_S in these two coordinates is $\boldsymbol{F}_R = \boldsymbol{B}_S^{\mathrm{T}}\boldsymbol{F}_S\boldsymbol{B}_S$. Since $\boldsymbol{B}_S$ is of dimension $3N$–$6\times m$, we have to adopt the following transformation to deduce $\boldsymbol{F}_S$ from $\boldsymbol{F}_R$.

By $\boldsymbol{F}_R = \boldsymbol{B}_S^{\mathrm{T}}\boldsymbol{F}_S\boldsymbol{B}_S$, we have $\boldsymbol{B}_S\boldsymbol{F}_R\boldsymbol{B}_S^{\mathrm{T}} = (\boldsymbol{B}_S\boldsymbol{B}_S^{\mathrm{T}})\boldsymbol{F}_S(\boldsymbol{B}_S\boldsymbol{B}_S^{\mathrm{T}})$. Now, $\boldsymbol{B}_S\boldsymbol{B}_S^{\mathrm{T}}$ is a symmetric square matrix of dimension $3N$–6 and its inverse is also symmetric. We then have $\boldsymbol{F}_S = (\boldsymbol{B}_S\boldsymbol{B}_S^{\mathrm{T}})^{-1}\boldsymbol{B}_S\boldsymbol{F}_R\boldsymbol{B}_S^{\mathrm{T}}(\boldsymbol{B}_S\boldsymbol{B}_S^{\mathrm{T}})^{-1}$ or $\boldsymbol{F}_S = [\boldsymbol{B}_S^{\mathrm{T}}(\boldsymbol{B}_S\boldsymbol{B}_S^{\mathrm{T}})^{-1}]^{\mathrm{T}}\boldsymbol{F}_R[\boldsymbol{B}_S^{\mathrm{T}}(\boldsymbol{B}_S\boldsymbol{B}_S^{\mathrm{T}})^{-1}]$.

The potential energy distribution of the coordinate i in mode k is $L^2{}_{ik}F_{ii}/\omega_k^2$.

3.5 Resonance

As the energies E_i^0, E_j^0 of the two states $|i\rangle$ and $|j\rangle$ are close to each other, their wave functions ψ_i^0, ψ_j^0 can mix up due to perturbation Hamiltonian H'. For instance, for

$$H' = \left(\frac{\partial^3 V}{\partial Q_i \partial Q_j^2}\right)_0 Q_i Q_j^2$$

the energies of the two new states are as follows:

$$E_+ = \frac{E_i^0 + E_j^0}{2} + \left[H_{ij}'^2 + \left(\frac{E_i^0 - E_j^0}{2}\right)^2\right]^{1/2}$$

$$E_- = \frac{E_i^0 + E_j^0}{2} - \left[H_{ij}'^2 + \left(\frac{E_i^0 - E_j^0}{2}\right)^2\right]^{1/2}$$

The corresponding wave functions are as follows:

$$\psi_+ = a\psi_i^0 + b\psi_j^0$$
$$\psi_- = b\psi_i^0 - a\psi_j^0$$

A common case is the Fermi resonance. For NCO^-, its two states ν_1^0 and $2\nu_2^0$ are close to each other. They can couple with each other and form two new states at $\nu_+ = 1298\text{cm}^{-1}, \nu_- = 1211\text{cm}^{-1}$. Meanwhile, they possess roughly equal spectral intensities (IR absorption) as shown in Fig. 3.3 [2].

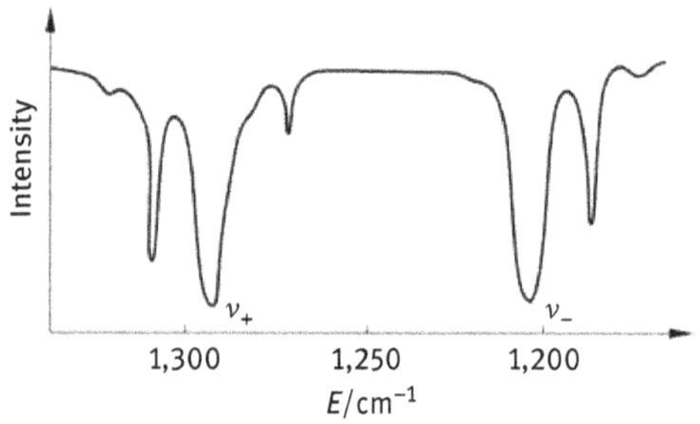

Fig. 3.3: By Fermi resonance, ν_1^0 and $2\nu_2^0$ couple to form ν_+ and ν_- in NCO^-.

3.6 Molecules with multiple stable configurations

Some molecules possess multiple stable configurations. Their potentials V are rotational angle α dependent as shown in Fig. 3.4.

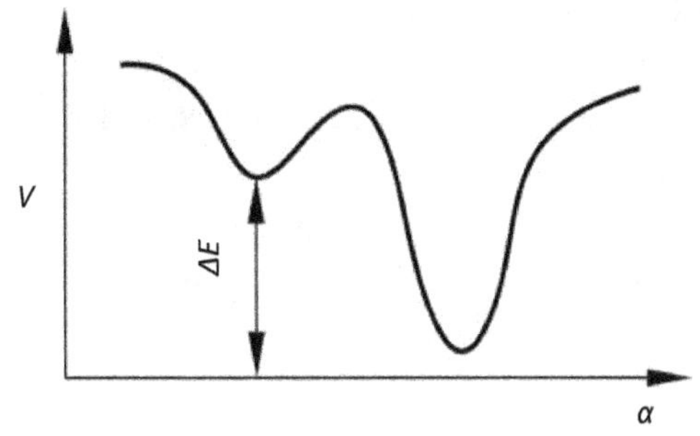

Fig. 3.4: The potential as a function of rotational angle *α*.

The properties of molecules with multiple stable configurations depend on the relative magnitudes of potential barrier and the thermal kinetic energy kT.

1. Potential barrier $\gg kT$

In this case, the transition among the various configurations is very difficult. Then we have different stable isomers as shown in Fig. 3.5.

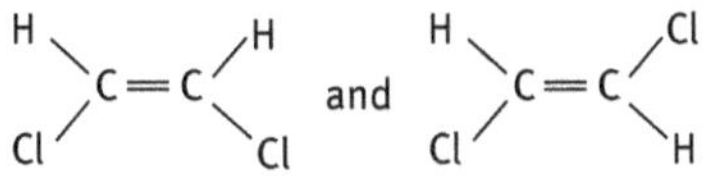

Fig. 3.5: Stable isomers of dichloroethene.

2. Potential barrier $\cong kT$

In this case, the molecule will transform among the various configurations. Their populations depend on $e^{-\Delta E/kT}$ as shown in Fig. 3.6. ΔE is their energy difference.

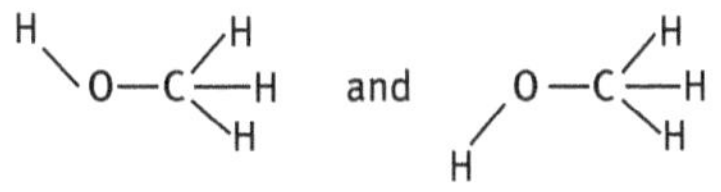

Fig. 3.6: The OH internal rotation of methanol.

3. Potential barrier $\ll kT$

In this case, the molecule can have free internal rotation as shown in Fig. 3.7; the two methyl groups of butyne can rotate freely around the molecular backbone.

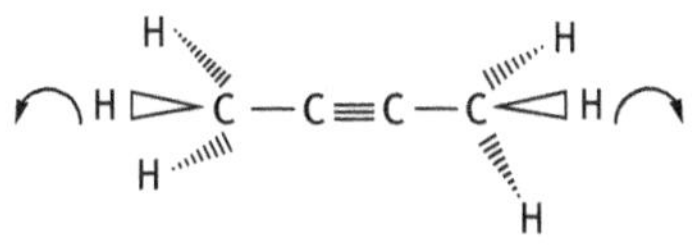

Fig. 3.7: The two methyl groups of butyne can rotate freely around the molecular backbone.

3.7 Molecular internal rotation

For the case of n stable configurations, potential energy $V(\alpha)$ can be expanded as follows:

$$V(\alpha) = a_0 + \sum_{j=1}^{\infty} a_j \cos jn\alpha$$

Taking α=0 as a stable point

$$V(0) = 0 = a_0 + \sum_{j=1}^{\infty} a_j$$

or

$$a_0 = -\sum_{j=1}^{\infty} a_j$$

then

$$V(\alpha) = -\sum_{j=1}^{\infty} a_j(1 - \cos jn\alpha)$$

For the first-order approximation,

$$V(\alpha) = -a_1(1 - \cos n\alpha)$$

As $\alpha = \pm\pi/n$, V attains its maximum V_0, that is,

$$V_0 = -a_1 \cdot 2$$

or

$$a_1 = -V_0/2$$

then

$$V(\alpha) = \frac{1}{2} V_0(1 - \cos n\alpha)$$

The Schrodinger equation is

$$-\frac{\hbar^2}{2I_r}\frac{d^2\psi(\alpha)}{d\alpha^2} + V(\alpha)\psi(\alpha) = E\psi(\alpha)$$

1. $V_0 \gg kT$

In this case, $\alpha \cong 0$. Expanding $V(\alpha)$ and neglecting the higher terms, we have the following:

$$
\begin{aligned}
V(\alpha) &= \tfrac{1}{2}V_0(1-\cos n\alpha)\\
&= \tfrac{1}{2}V_0\left[1-\left(1-\tfrac{n^2\alpha^2}{2!}+\tfrac{n^4\alpha^4}{4!}-\cdots+\cdots\right)\right]\\
&\approx \tfrac{1}{2}\left(\tfrac{V_0 n^2}{2}\right)\alpha^2
\end{aligned}
$$

This is of the functional form $\frac{1}{2}kq^2$. The eigenenergy is as follows:

$$
\begin{aligned}
E_V &= \hbar\omega_r\left(V+\frac{1}{2}\right)\\
&= \hbar\sqrt{\frac{V_0 n^2}{2I_r}}\left(V+\frac{1}{2}\right)\\
V &= 0, 1, 2, \cdots
\end{aligned}
$$

2. $V_0 \ll kT$

As $V_0 \ll kT$, the internal rotation is free. For $E \gg V_0$, $V(\alpha)$ can be treated as half its average value:

$$V(\alpha) = \frac{1}{2}V_0$$

The Schrodinger equation is as follows:

$$\frac{d^2}{d\alpha^2}\psi + \frac{2I_r}{\hbar^2}\left(E-\frac{V_0}{2}\right)\psi = 0$$

Let

$$\beta^2 = \frac{2I_r}{\hbar^2}\left(E-\frac{V_0}{2}\right)$$

then

$$\psi = e^{\pm i\beta\alpha}$$

By the periodic property,

$$\psi\left(\alpha+\frac{2\pi}{n}\right) = \psi(\alpha)$$

we have

$$e^{\pm i\beta\frac{2\pi}{n}} = 1$$

or $\beta = nJ$

with $J = 0, 1, 2, \cdots$

The energy is

$$E_J = \frac{1}{2}V_0 + \frac{\hbar^2}{2I_r}\beta^2$$
$$= \frac{1}{2}V_0 + \frac{\hbar^2 n^2}{2I_r}J^2$$

The angular momentum is $Jn\hbar$. The condition that $E_J \gg V_0$ means that J cannot be too small integer.

3. Potential barrier $\approx kT$
In this case, the low lying states can be treated as in the first situation, $V_0 \gg kT$. For the high lying states, the second situation, $V_0 \ll kT$, can be adopted. For general states, the situation is more complicated [3] and we need to solve the Schrodinger equation.

A molecule possessing multiple stable configurations usually have the same number of level degeneracy. The degeneracy often splits due to perturbation. For example, NH_3 possesses two identical stable configurations through inversion as shown in Fig. 3.8.

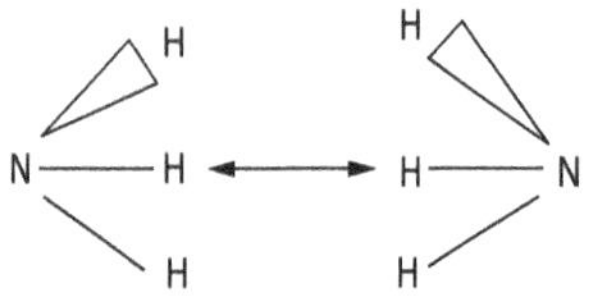

Fig. 3.8: The two identical stable configurations of NH_3 through inversion.

Its potential and levels are shown in Fig. 3.9. The splitting of degeneracy is obvious.

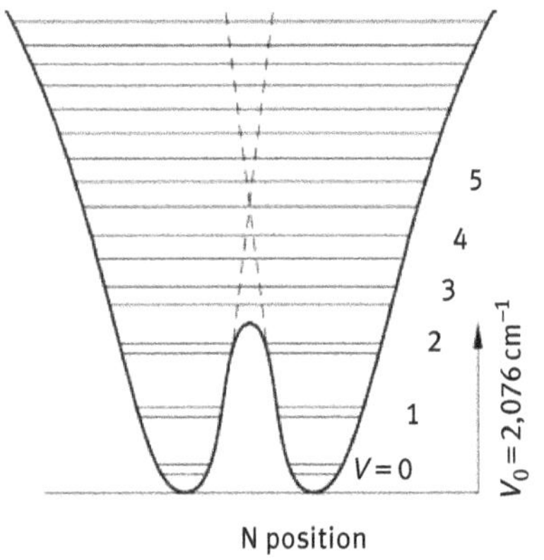

Fig. 3.9: The potential and levels of NH_3. Note the splitting of degeneracy.

3.8 Vibrational frequencies of functional groups

Often in a normal mode, one or few atoms possess much larger vibrational amplitudes. For example, in methanol, there is a mode like as shown in Fig. 3.10, where the amplitude of the O–H bond stretch is much more significant. Its wave number is 3,600 cm^{-1}. Other molecules possessing similar structures with the O–H bond may also have a mode due to O–H stretch, with wave number about 3,600 cm^{-1}. We therefore infer that the characteristic wave number of the O–H functional group is 3,600 cm^{-1}.

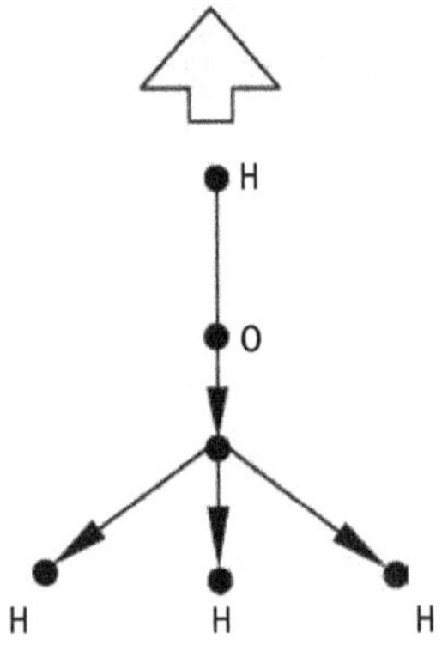

Fig. 3.10: In methanol, there is a mode for which the amplitude of the O–H bond stretch is much more significant.

Table 3.1 lists the characteristic wave numbers of some functional groups.

Tab. 3.1: The characteristic wave numbers of some functional groups.

Functional group	Wave number	Functional group	Wave number
−C≡N	2,100	›N−H	3,350
−C≡C−	2,050	→C−H	2,960
›C=C‹	1,650	→C−H	3,020
›C=O	1,700	≡C−H	3,300
−O−H	3,650		

By the characteristic wave numbers of functional groups, we often determine the molecular structure.

3.9 Remarks

The idea of normal mode is based on the harmonic approximation. This picture of molecular vibration is adequate for lying vibrational states. At high excitation, harmonic approximation is no longer adequate and the idea and the picture based on normal mode are no longer true. Then, the motion is close to the classical realm such that the nonlinear effect including chaos will emerge. Because of nonlinear effect, sometimes energy may aggregate/localize around a bond, like a soliton. However, this localization of energy is different from the motion of functional groups mentioned in Section 3.8. This is a new research frontier. The algorithm to explore its physics is entirely different from the treatment discussed in this chapter. We will discuss this issue in Chapter 13.

Exercises

3.1 For a molecule, there are two coordinate systems S_1 and S_2. Their relation is

$$\mathbf{A}\mathbf{S}_1 = \mathbf{S}_2$$

1. Show that their momenta, $\boldsymbol{P}_{S_1}$ and $\boldsymbol{P}_{S_2}$, are related by

$$A^{\mathrm{T}}\boldsymbol{P}_{S_2} = \boldsymbol{P}_{S_1}$$

2. In terms of S_1 and S_2, T and V are given as follows:

$$T_{S_1} = \tfrac{1}{2}\boldsymbol{P}_{S_1}^{\mathrm{T}}\boldsymbol{G}_{S_1}\boldsymbol{P}_{S_1},\ T_{S_2} = \tfrac{1}{2}\boldsymbol{P}_{S_2}^{\mathrm{T}}\boldsymbol{G}_{S_2}\boldsymbol{P}_{S_2},$$
$$V_{S_1} = \tfrac{1}{2}\boldsymbol{S}_1^{\mathrm{T}}\boldsymbol{F}_{S_1}\boldsymbol{S}_{S_1},\ V_{S_2} = \tfrac{1}{2}\boldsymbol{S}_2^{\mathrm{T}}\boldsymbol{F}_{S_2}\boldsymbol{S}_2$$

Furthermore, we have

$$\boldsymbol{G}_{S_2} = \boldsymbol{A}\boldsymbol{G}_{S_1}\boldsymbol{A}^{\mathrm{T}}$$
$$\boldsymbol{F}_{S_2} = \left(\boldsymbol{A}^{\mathrm{T}}\right)^{-1}\boldsymbol{F}_{S_1}\boldsymbol{A}^{-1}$$
$$\boldsymbol{G}_{S_2}\boldsymbol{F}_{S_2} = \boldsymbol{A}\boldsymbol{G}_{S_1}\boldsymbol{F}_{S_1}\boldsymbol{A}^{-1}$$

and

$$\boldsymbol{L}_{S_2} = \boldsymbol{A}\boldsymbol{L}_{S_1}$$

3. When is $\boldsymbol{A}^{\mathrm{T}}=\mathbf{A}^{-1}$?, show that, in Section 3.2, $\boldsymbol{L}^{-1}$ possesses the following property:

$$\left(\boldsymbol{L}^{-1}\right)_{ki} = L_{ik}$$

3.2 Derive the expressions of E_+, E_- in Section 3.5.

3.3 For the system shown in Fig. 3.11, the internal coordinates r_1, r_2 and β are defined therein.

Defining S_1, S_4, S_5 as follows in Fig.3.11:

$$S_4 = \frac{1}{\sqrt{2}}(r_1 + r_2)$$
$$S_5 = \rho\beta/2$$
$$S_1 = \frac{1}{\sqrt{2}}(r_1 - r_2)$$

Here, ρ is the equilibrium distance of C–O, and μ_1, μ_Y, μ_Z are its dipole and the components along y and z directions, respectively.

Infer the following relations:

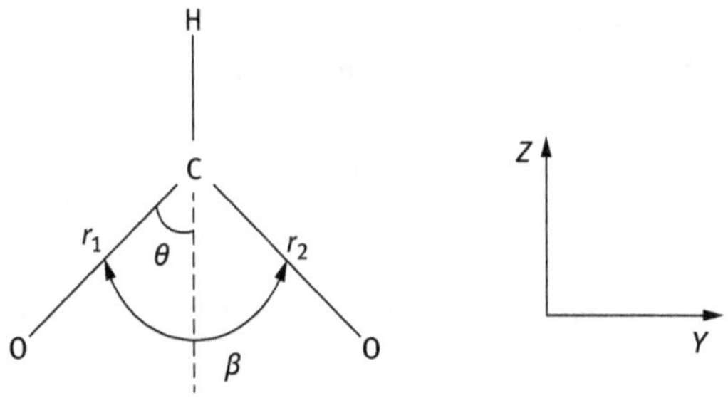

Fig. 3.11: The structure for Exercise 3.3.

(1) $\frac{\partial \mu_1}{\partial r_1} = \frac{\mu_1}{\rho} = \frac{1}{\cos\theta}\frac{1}{\sqrt{2}}\frac{\partial \mu_Z}{\partial S_4}$ (2) $\frac{\partial \mu_Y}{\partial S_1} = \frac{\partial \mu_Z}{\partial S_4}\tan\theta$

(3) $\frac{\partial \mu_Z}{\partial S_5} \cong \frac{-\mu_1}{\rho}\sin\left(\theta + \frac{\beta}{2}\right)$ (4) $\frac{\partial S_5}{\partial S_4} = -\frac{\cos\theta\sqrt{2}}{\sin(\theta+\beta/2)}$

(5) $\frac{\partial \mu_Z}{\partial S_4} = \frac{\partial \mu_Z}{\partial S_5}\frac{-\cos\theta\sqrt{2}}{\sin(\theta+\beta/2)}$

References

[1] Wilson EB, Decius JC, Cross PC. Molecular Vibrations. New York: McGraw Hill, 1955.
[2] Decius JC, Gordon DJ. J Chem Phys, 1967, 47: 1286.
[3] Wilson EB. Chem Revs, 1940, 27: 17.

4 Force constants of thiocyanate ion adsorbed on electrode surface

4.1 Introduction

In Chapter 3, we have discussed the normal mode analysis. In this analysis, as the structure of a molecule is known, the ***G*** matrix can be easily obtained from the bond lengths, angles and atomic masses. Since the number of normal mode frequencies, which can be obtained from the spectra is less than the number of the force-field elements, the retrieval of the force-field matrix seems impossible.

However, the situation is not so worse. One can first guess the force constants based on the chemical/physical aspects and calculate the frequencies. From the deviation of the calculated and observed frequencies, the force constants can be adjusted so that the calculated frequencies are fitted to the observed frequencies. In this way, after several rounds of fitting, the final force field can be nailed down. For a bond species present in different molecules, their force constants cannot be identical, but the variation should not be too large. This helps in the determination of force field. The theoretical force field by quantum chemical calculation is also a good reference for its determination. In the fitting process, sometimes, only few force constants, instead of all, are adjusted at a particular time. This also simplifies the process.

Currently, software for the normal mode analysis are popular. The collection of the force constant data in the past decades is immense. The determination of force field has become an easy task now.

In this chapter, the force constant will be determined and emphasis will be on how can it be beneficial in the analysis of the spectroscopic observation.

4.2 Vibrational analysis of thiocyanate ion adsorbed on the Ag electrode surface

As thiocyanate ion, SCN^-, is adsorbed on the surface of the Ag electrode, its Raman intensity will be enhanced up to an order of 10^6[1]. This is called the surface-enhanced Raman scattering. It is an interesting phenomenon; we will address this in Chapter 9.

The vibrational frequencies of the adsorbed SCN^- will be affected by the electrode potential. Because of adsorption, a new bond between the S atom and the surface is formed. The system possesses three stretching modes. Their potential-dependent frequencies are listed in Tab. 4.1. As the voltage shifts from −0.2 to −0.8 V (with reference to the standard Calomel electrode [SCE]), the frequencies of υ_1, υ_3 modes decrease, while that of υ_2 mode increases. As the voltage is −0.8 V, the

https://doi.org/10.1515/9783110625097-004

Tab. 4.1: The potential-dependent mode frequencies of SCN^- adsorbed on the Ag electrode surface.

V/V_{SCE}	υ_1 (cm^{-1})	υ_2(cm^{-1})	υ_3(cm^{-1})
In solution	–	745	2,072
−0.8	190	740	2,085
−0.7	196	738	2,092
−0.6	203	736	2,096
−0.5	210	733	2,100
−0.4	216	727	2,107
−0.3	220	716	2,112
−0.2	225	712	2,117

frequencies of υ_2, υ_3 modes are close to those in the solution. The adsorption is then very weak.

For the normal mode analysis of this system, we first construct an adsorption configuration as shown in Fig. 4.1, where the mass M of the surface is variable. If SCN^- is adsorbed directly on the surface, M is *infinite*. If SCN^- is adsorbed on a single Ag atom (called adatom), which is detached and has no effect from the surface, then M is the mass of the Ag atom. Otherwise, because of the surface effect, M will be smaller than the Ag mass (caused by the charge repulsion between the adatom and the bulk surface). For convenience, we set M as cM_{Ag}, where M_{Ag} is the mass of the Ag atom. Corresponding to these three situations, c is ∞, 1 and <1.

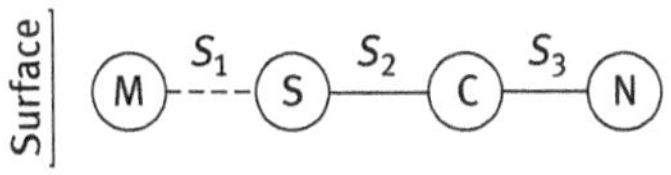

Fig. 4.1: The SCN^- adsorption configuration. S_1, S_2, S_3 are the stretching coordinates.

The analysis shows that $c=\infty$ or 1 cannot lead to a consistent result. However, as $c=0.1$, the fitting is very good. The results are listed in Tab. 4.2. Roughly, the force constant K_{CN} of C–N is three times that of the S–C bond, K_{SC}. This is consistent with the recognition that C≡N is a triple bond and S–C is a single bond. K_{SC} in solution is larger than under adsorption. K_{MS} is very small; it is only 1/20 of K_{SC}. Hence, the M–S bonding is rather weak. The adsorption is physical and not a chemical bonding. In this table, the sensitive behavior of the force constants on the applied potential is observed. As the potential shifts from −0.2 to −0.8 V, K_{MS} decreases, showing that the adsorption becomes weaker. Then, K_{SC} increases while K_{CN} decreases. The ratio of K_{CN}/K_{SC} drops from 3.34 to 2.98, which is close to the solution case, 2.76. In this case, C≡N is more sensitive to the applied electrode potential. This may be

Tab. 4.2: The force constants (10^{-3} dyn/ Å) of M–SCN$^-$, where the correcting factor is $c = 0.1$. V_{SCE} is the standard calomel electrode.

V/V_{SCE}	K_{MS}	K_{SC}	K_{CN}
Solution	–	5.28	14.57
−0.8	0.195	5.13	14.84
−0.7	0.208	5.09	14.96
−0.6	0.224	5.05	15.04
−0.5	0.239	4.99	15.13
−0.4	0.253	4.89	15.29
−0.3	0.263	4.72	15.43
−0.2	0.275	4.65	15.54

attributed to the fact that it is farther away from the surface and possesses more electrons.

From this example, we know that the normal mode analysis is quite routine. The important point is how to apply it appropriately to get significant results.

Reference

[1] Huang Y, Wu G. Spectrochimica Acta, 1989, 45A:123.

5 Representations of point groups and their applications (thiocynante)

5.1 Molecular symmetry and the definition of groups

Many molecules possess symmetries. This chapter discusses about the molecular symmetry that refers to the geometric invariance of a molecule under a spatial transformation. Thus, we say that the molecule is symmetric under the transformation, which is also called the symmetry operation. The invariants under operations such as point, axis and plane (on which the operations are based) are called symmetry elements. Each symmetry operation corresponds to an element as shown in Tab. 5.1. The reflection mirrors can be further classified as σ_v, σ_h and σ_d, if the mirror is vertical to, is in a plane or bisects a dihedral angle, respectively. These operations are shown in Fig. 5.1.

Tab. 5.1: Symmetry elements and operations.

Element	Notation	Operation	Notation
Identity	E	No action	E
Mirror	σ	Reflection	σ
Center	i	Inversion	i
Axis	C_n	Rotation of $2\pi k/n$, $k = 1, 2, 3, \ldots$ along an axis	C_n^k
Skew axis	S_n	Rotation of $2\pi k/n$, $k = 1, 2, 3, \ldots$ along an axis Followed by reflection through a mirror vertical to the axis	S_n^k

It is easy to confirm that:

1. For two symmetry operations, A and B, their combined operation, denoted as $A{\cdot}B$, that is, B operation is followed by A, is also called a symmetry operation. This is the closure property.
2. For an operation A, there is another operation B such that $A{\cdot}B = B{\cdot}A = E$. B is called the inverse of A.
3. For operations A, B, C, it is true that $A{\cdot}(B{\cdot}C) = (A{\cdot}B){\cdot}C$. This is the associative property.
4. Identity, E, is always a symmetry operation.

The collection of symmetry operations (there is at least E) satisfies the above four properties. Mathematically, we say that they form a group, or a symmetric group.

https://doi.org/10.1515/9783110625097-005

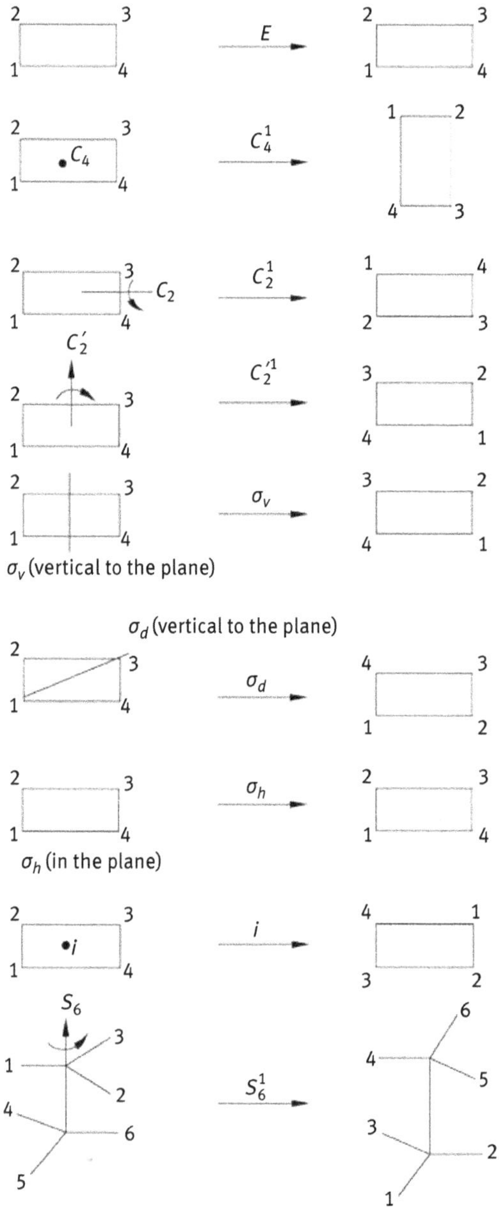

Fig. 5.1: The symmetry operations.

There are many sets that form groups. For example, the set of integers forms a group under addition, since

1. $n_1 + n_2 = n_3$
2. $n_1 + (-n_1) = 0$ (0 is the identity)
3. $n_1 + (n_2 + n_3) = (n_1 + n_2) + n_3$
4. $n_1 + 0 = 0 + n_1 = n_1$

The molecular symmetry group is to classify the vibrational and electronic states. In other words, the wave functions of vibrational and electronic states will possess specific transformational properties under the symmetry group. Even the details of these motions are not explicitly known, and their transformational properties can be unequivocally known. The algorithm of this analysis is very simple. This is the reason that we need to study the symmetry groups.

Why do the wave functions of vibrational and electronic states possess specific transformational properties under the symmetry group? We start with the Schrödinger equation:

$$H\psi = E\psi$$

where H is the system Hamiltonian, ψ and E are the wave function and its eigenenergy. Suppose that R is a symmetry operation, then

$$RHR^{-1}R\psi = ER\psi$$

Since H is invariant under R, we have

$$HR\psi = ER\psi$$

In other words, $R\psi$ and ψ possess the same energy E. One possibility is that $R\psi$ and ψ are different and are degenerate states. Suppose that this is not the case, then $R\psi$ and ψ can only differ by a phase factor, that is,

$$R\psi = e^{i\theta}\psi$$

The simplest situation is that $e^{i\theta}$ is +1 or −1, then

$$R\psi = \pm\psi$$

In other words, the eigenvalue of ψ under R is 1 or −1. This example demonstrates that the vibrational and electronic wave functions will possess specific transformational properties under the symmetry group. This is a very generic and important property.

5.2 Classification of groups

Groups can be classified as follows:

1. Finite group: This is a group whose number of elements, called the order, is finite.
2. Infinite group: This is a group whose number of elements is not finite.
3. Abelian group: This is a group that its elements A and B commute
$$A \cdot B = B \cdot A$$
4. Cyclic group: This is a group whose elements can be expressed as $X^n (n = 0, 1, 2, \ldots)$ by an element X.
5. Subgroup: The subset F of a group G that forms a group. F is called the subgroup of G.

5.3 Some properties of groups

1. Coset

Suppose that G contains elements $\{S_1 \equiv E, S_2, \ldots, S_g\}$ and some of its elements $\{S_1, S_2, \ldots, S_s\}$ form a subgroup F, then we can choose X that is not in F to form the set XF by

$$X \begin{bmatrix} E \\ \vdots \\ S_S \end{bmatrix} = \begin{bmatrix} X \\ \vdots \\ XS_S \end{bmatrix} \equiv XF$$

or

$$\begin{bmatrix} E \\ \vdots \\ S_S \end{bmatrix} X = \begin{bmatrix} X \\ \vdots \\ S_S X \end{bmatrix} \equiv FX$$

The former is called the left coset and the latter the right coset.

It is easy to show that for two cosets XF and YF, if they share an element, then they are identical. Hence, the elements of a group G can be distributed in the cosets without intersections:

$$\begin{aligned} \{E, S_2, \ldots, S_g\} = \{&E, S_2, \ldots, S_s\} \\ &+ \{S_{S+1}, \ldots, S_{2s}\} \\ &+ \{\cdots\} + \cdots \end{aligned}$$

For this decomposition, there are l cosets. Each coset possesses the same number of elements. Hence,

$$g = s \cdot l$$

That is, the order of a group G (the number of elements) is the multiple of the order of its subgroup.

2. Class

For two elements A and B, if there is another element X such that

$$B = XAX^{-1}$$

then A and B are conjugates to each other. The set of elements that are conjugate to each other forms a class. If X is a coordinate transformation, then the elements in a class are identical in essence that their difference is bare as they are in different coordinate systems.

C is employed for the notation of class. Note that E forms a class by itself.

3. Invariant subgroup and quotient group

Suppose that a group G contains a subgroup F. If for all the elements X of G, we have

$$XFX^{-1} = F$$

or

$$XF = FX$$

that is, the right and left cosets of F are the same; then F is called an invariant subgroup.

We can distribute the elements of G into cosets by F:

$$\begin{aligned}\{E, S_2, \ldots, S_g\} = \{E, S_2, \ldots, S_s\} \\ + \{S_{s+1}, \ldots, S_{2s}\} \\ + \{\cdots\} + \cdots\end{aligned}$$

Now, consider each coset as an element, $K_i(\equiv FK_i)$, then

$$\begin{aligned}K_iK_j &\equiv (FK_i)(FK_j) \\ &= (K_iF)(FK_j) \\ &= K_iFK_j = FK_iK_j \\ &= FK_k = K_k\end{aligned}$$

Meanwhile, we have

$$F \cdot K_i = K_i$$

$$K_i K_i^{-1} = F$$

$$(K_i K_j) K_k = K_i (K_j K_k)$$

Hence, $\{F, K_1, \ldots, K_{f-1}\}$ forms a group, which is called the quotient group of G with respect to the invariant subgroup F.

4. Homomorphic and isomorphic groups

Two groups "H and K" are called homomorphic, if there exists a correspondence:

$$\{h_{i1}, h_{i2}, \ldots, h_{ir}\} \leftrightarrow K_i$$

such that if

$$h_{im} \cdot h_{jn} = h_{kp}$$

then

$$K_i K_j = K_k$$

also holds in K. If the correspondence is one to one, then H and K are called isomorphic.

A group and its quotient group are homomorphic.

5.4 Point groups

The set of symmetry operations of a molecule forms a group. Since, for such a group, there is always a point that remains fixed, it is called a point group. There are mainly 32 point groups.

They can be classified as follows:

1. The group formed by E, denoted by C_1
2. The group formed by E and σ_h, denoted by C_s
3. The group formed by E and i, denoted by C_i
4. The group formed by E, C_n^i $(i = 1, \ldots, n-1)$, denoted by C_n
5. The group formed by the elements of C_n and σ_h, denoted by C_{nh}
6. The group formed by the elements of C_n and σ_v, denoted by C_{nv}
7. The group formed by the elements of C_n and C_2 vertical to C_n, denoted by D_n
8. The group formed by the elements of D_n and σ_d, denoted by D_{nd}
9. The group formed by the elements of D_n and σ_h, denoted by D_{nh}

10. The group formed by E, S_n^i $(i=1, \ldots, n-1)$, denoted by S_n (if n is odd, then $S_n = C_{nh}$)
11. Other cubic groups such as T, T_d, O and O_h

5.5 Group representation

If there is homomorphic correspondence between the element A of group G and $n \times n$ unitary matrix (a matrix is unitary if its transposed, complex conjugate matrix is its inverse) $\mathbf{\Gamma}(A)$:

$$A \rightarrow \mathbf{\Gamma}(A)$$

with the property that if $AB = C$, then

$$\mathbf{\Gamma}(A)\mathbf{\Gamma}(B) = \mathbf{\Gamma}(C)$$

The set of matrices $\mathbf{\Gamma}(A)$ is called the representation of G. Or, we simply say that $\mathbf{\Gamma}(A)$ is a representation.

The elements of a group are, more or less, abstract. Via representation, we can manipulate the group operation by matrix multiplication. This is very convenient.

The representation of a group is not unique. Suppose that there is $\boldsymbol{S}$ such that

$$\mathbf{\Gamma}'(A) = \boldsymbol{S}^{-1}\mathbf{\Gamma}(A)\boldsymbol{S}$$

Hence, if $AB = C$

then

$$\begin{aligned}\mathbf{\Gamma}'(A)\mathbf{\Gamma}'(B) &= \boldsymbol{S}^{-1}\mathbf{\Gamma}(A)\boldsymbol{S}\boldsymbol{S}^{-1}\mathbf{\Gamma}(B)\boldsymbol{S}\\ &= \boldsymbol{S}^{-1}\mathbf{\Gamma}(A)\mathbf{\Gamma}(B)\boldsymbol{S}\\ &= \boldsymbol{S}^{-1}\mathbf{\Gamma}(C)\boldsymbol{S}\\ &= \mathbf{\Gamma}'(C)\end{aligned}$$

In other words, $\mathbf{\Gamma}'(A)$ is also a representation of G. We say that $\mathbf{\Gamma}(A)$ and $\mathbf{\Gamma}'(A)$ are equivalent. The transformation

$$\mathbf{\Gamma}'(A) = \boldsymbol{S}^{-1}\mathbf{\Gamma}(A)\boldsymbol{S}$$

is called the similarity transformation.

If a group representation $\mathbf{\Gamma}^{(r)}(A)$ can be transformed by a similarity transformation to

$$\mathbf{\Gamma}^{(r)}(A) = \begin{bmatrix} \mathbf{\Gamma}^{(1)}(A) & 0 \\ 0 & \mathbf{\Gamma}^{(2)}(A) \end{bmatrix}$$

then $\mathbf{\Gamma}^{(r)}$ is reducible. Of course, $\mathbf{\Gamma}^{(1)}$ and $\mathbf{\Gamma}^{(2)}$ are also representations. If $\mathbf{\Gamma}^{(r)}(A)$ cannot be transformed to the block-diagonal form, then $\mathbf{\Gamma}^{(r)}$ is irreducible.

Two nonequivalent, irreducible and unitary representations $\mathbf{\Gamma}^{(i)}(R)$ and $\mathbf{\Gamma}^{(j)}(R)$ possess the following orthogonality relations:

$$\sum_{R\in G}[\mathbf{\Gamma}^{(i)}(R)]^{*}_{uv}[\mathbf{\Gamma}^{(j)}(R)]_{\alpha\beta} = \frac{g}{d_i}\delta_{ij}\delta_{\mu\alpha}\delta_{v\beta}$$

where R is the group element, d_i is the dimension of the irreducible representation $\mathbf{\Gamma}^{(i)}(R)$, g is the group order and α, β, μ, v are indices denoting the positions in a matrix.

5.6 Character

The sum of the diagonal elements of the representation matrix is called the character, denoted by $\chi^{(i)}(R)$:

$$\chi^{(i)}(R) = \sum_{\mu}[\mathbf{\Gamma}^{(i)}(R)]_{\mu\mu}$$

The characters of equivalent representations are the same since two matrices are related by a similarity transformation:

$$B = XAX^{-1}$$

share the same trace, or character.

The elements of a class share the same character. In contrast to the representation matrices, characters are simple numbers. Their operations are simple. The character of the identity operation is the dimension of its representation, that is, $\chi^{(i)}(E) = d_i$.

5.7 Character tables

The characters of the irreducible representations of a group are tabulated in Tab. 5.2, called the character table.

In Tab. 5.2, C_i shows the class of G, χ is the character and Γ is the labeling of the irreducible representations. A character table possesses the following important properties:

1. The number of irreducible representations and the number of classes are the same. Hence, a character table is square.
2. $\sum_{k}\chi^{(i)}(C_k)^{*}\chi^{(j)}(C_k)N_k = g\delta_{ij}$

Tab. 5.2: The character table of a group.

G	E	C_1	$C_2 \dots$
$\Gamma^{(1)}$	$\chi^{(1)}(E)$	$\chi^{(1)}(C_1)$	$\chi^{(1)}(C_2)$
$\Gamma^{(2)}$	$\chi^{(2)}(E)$	$\vdots$	$\vdots$
$\vdots$	$\vdots$	$\vdots$	$\vdots$

where N_k is the number of elements in class k and $\chi^{(i)}(C_k)$ is its character.

This is because that

$$
\begin{aligned}
&\sum_k \chi^{(i)}(C_k)^* \chi^{(j)}(C_k) N_k \\
&= \sum_R \chi^{(i)}(R)^* \chi^{(j)}(R) \\
&= \sum_R \left[\sum_\mu \mathbf{\Gamma}^{(i)}(R)_{\mu\mu} \right]^* \left[\sum_\alpha \mathbf{\Gamma}^{(j)}(R)_{\alpha\alpha} \right] \\
&= \sum_{\mu\alpha} \sum_R \left[\mathbf{\Gamma}^{(i)}(R) \right]^*_{\mu\mu} \left[\mathbf{\Gamma}^{(j)}(R) \right]_{\alpha\alpha} \\
&= \sum_{\mu\alpha} \frac{g}{d_i} \delta_{ij} \delta_{\mu\alpha} \delta_{\mu\alpha} \\
&= \frac{g}{d_i} \delta_{ij} d_i \\
&= g\delta_{ij}
\end{aligned}
$$

These two properties show that if the characters are viewed as row or column vectors, they are orthogonal and normalized to g.

3.

$$
\sum_i \chi^{(i)}(C_k)^\star \chi^{(i)}(C_l) = \frac{g}{N_k} \delta_{kl} \tag{5.1}
$$

4. The sum of squares of the dimensions ($\chi^{(i)}(E)$) of the irreducible representations is equal to the order of a group, that is,

$$
\sum_i d_i^2 = g \tag{5.2}
$$

This is because that $\chi^{(i)}(E) = d_i$, by eq. (5.1) and set $C_k = E$, then we have eq. (5.2).

Irreducible representations are labeled by the following notations:

1. One-dimensional irreducible representation is labeled by A, B. A is for $\chi(C_n) = +1$ and B is for $\chi(C_n) = -1$. C_n is the principal axis.
2. Two and three dimensions are labeled by E and T (or F), respectively.
3. Subscripts 1 and 2 are for $\chi(\sigma_v)$ or $\chi(C_2)$ (C_2 is vertical to the principal axis) being + or −, respectively.
   Superscripts ′ and ″ are for $\chi(\sigma_h)$ being + or −, respectively.
   Subscripts g and u are for $\chi(i)$ being + or −, respectively.

5.8 Reduction of a representation

Suppose a matrix representation $\mathbf{\Gamma}(R)$ is reduced to

$$\begin{bmatrix} \mathbf{\Gamma}^{(1)}(R) & 0 & 0 \\ 0 & \mathbf{\Gamma}^{(2)}(R) & 0 \\ 0 & 0 & \ddots \end{bmatrix}$$

where $\mathbf{\Gamma}^{(i)}(R)$s are irreducible representations, then we have

$$\chi(R) = \sum_j a_j \chi^{(j)}(R)$$

where a_j is the number of $\mathbf{\Gamma}^{(j)}(R)$ that appears in the reduction.

Since

$$\begin{aligned} & \sum_R \chi(R)\chi^{(i)}(R)^* \\ &= \sum_R \sum_j a_j \chi^{(j)}(R)\chi^{(i)}(R)^* \\ &= \sum_j a_j \sum_R \chi^{(j)}(R)\chi^{(i)}(R)^* \\ &= \sum_j a_j g \delta_{ij} \\ &= a_i g \end{aligned}$$

then

$$a_i = \frac{1}{g} \sum_R \chi(R)\chi^{(i)}(R)^* \tag{5.3}$$

or

$$a_i = \frac{1}{g} \sum_k \chi(C_k)\chi^{(i)}(C_k)^* N_k \tag{5.4}$$

This formula offers us an algorithm to obtain the number of the irreducible representation $\Gamma^{(i)}$ contained in a representation.

5.9 Basis

Suppose that a set of coordinates ξ (or functions) is changed to ξ' by the action of R. In matrix notation, we have

$$\begin{bmatrix} \xi'_1 \\ \xi'_2 \\ \vdots \end{bmatrix} = \begin{bmatrix} R_{11} & R_{12} & \cdots \\ \vdots & \cdots & \cdots \\ \vdots & & \end{bmatrix} \begin{bmatrix} \xi_1 \\ \xi_2 \\ \vdots \end{bmatrix}$$

or simply,

$$\boldsymbol{\xi}' = \boldsymbol{R}\boldsymbol{\xi}$$

Since $\boldsymbol{R}$ is the representation of R, we say that ξ is the basis of $\boldsymbol{R}$. For instance, if

$$\begin{bmatrix} x \\ y \\ z \end{bmatrix}$$

is employed as the basis, the representation of i is

$$\begin{bmatrix} -1 & 0 & 0 \\ 0 & -1 & 0 \\ 0 & 0 & -1 \end{bmatrix}$$

and $C_z(\theta)$ is

$$\begin{bmatrix} \cos\theta & -\sin\theta & 0 \\ \sin\theta & \cos\theta & 0 \\ 0 & 0 & 1 \end{bmatrix}$$

Suppose under the similarity transformation by $\boldsymbol{S}$, $\boldsymbol{R}$ is transformed to

$$\boldsymbol{SRS}^{-1} = \begin{bmatrix} D_1 & & & & \\ & D_2 & & & \\ & & D_3 & & \\ & & & \ddots & \\ & & & & D_i \end{bmatrix}_R$$

meanwhile,

$$\mathbf{S}\xi = \begin{bmatrix} \xi''_1 \\ \xi''_2 \\ \vdots \\ \xi''_i \end{bmatrix}$$

then (ξ''_k) is the basis of D_k.

Terminologically, for the basis of an irreducible representation, it is often said that "the basis (coordinates or functions) belongs to that irreducible representation" or that "the basis (coordinates or functions) is the symmetry of that irreducible representation." This is because that the characters of an irreducible representation show the transformation properties or the symmetries of its basis under the symmetry operations. For instance, in D_2 group, z under $C_2(x)$, $C_2(y)$, $C_2(z)$ is changed to $-z$, $-z$, z. The characters (−1, −1, 1) show the symmetric properties of z in D_2.

The bases of irreducible representations are shown in the character tables.

The character tables of the point groups are shown in Appendix. In the character table, besides the group elements (classes) and the irreducible representations, we have to pay attention to the symmetries (or irreducible representations) to which x, y, z, xx, xy, xz, yz and R_x, R_y, R_z (rotations along x, y and z axes) belong.

5.10 Normal coordinates as the bases

The potential of molecular vibration $V(Q)$ is written as

$$V(Q) = \frac{1}{2}\sum_k \omega_k^2 \sum_\alpha Q_{k\alpha}^2$$

α is for the degenerate states corresponding to mode k.

Under the symmetry operation $\mathbf{S}$, we have $(\mathbf{Q} = \mathbf{S}^{-1}\mathbf{Q}')$

$$Q_{k\alpha} = \sum_l \sum_\beta S^{-1}{}_{k\alpha,\, l\beta} Q'_{l\beta}$$

Hence,

$$V(Q') = \frac{1}{2}\sum_k \omega_k^2 \sum_\alpha \left[\sum_{l,\beta} S^{-1}{}_{k\alpha,\, l\beta} Q'_{l\beta}\right]\left[\sum_{m,\gamma} S^{-1}{}_{k\alpha,\, m\gamma} Q'_{m\gamma}\right]$$

$$= \frac{1}{2}\sum_{l\beta}\sum_{m\gamma}\left[\sum_{k\alpha} \omega_k^2 S^{-1}{}_{k\alpha,\, l\beta} S^{-1}{}_{k\alpha,\, m\gamma}\right] Q'_{l\beta} Q'_{m\gamma}$$

Since

$$V(Q') = \frac{1}{2}\sum_{l\beta} \omega_l^2 Q'^2_{l\beta}$$

then

$$\sum_{k\alpha} \omega_k^2 \, S^{-1}{}_{k\alpha,\, l\beta} S^{-1}{}_{k\alpha,\, mr} = \omega_l^2 \delta_{ml} \delta_{\beta\gamma}$$

Multiply this expression by $S^{-1}{}_{n\delta,\, m\gamma}$ and take sum over m, γ, then we have

$$\sum_{m\gamma} \sum_{k\alpha} \omega_k^2 S^{-1}{}_{k\alpha,\, l\beta} S^{-1}{}_{k\alpha,\, m\gamma} S^{-1}{}_{n\delta,\, m\gamma} = \sum_{m\gamma} \omega_l^2 S^{-1}{}_{n\delta,\, m\gamma} \delta_{ml} \delta_{\beta\gamma}$$

$$\sum_{k\alpha} \omega_k^2 S^{-1}{}_{k\alpha,\, l\beta} \sum_{m\gamma} S^{-1}{}_{k\alpha,\, m\gamma} S^{-1}{}_{n\delta,\, m\gamma} = \omega_l^2 S^{-1}{}_{n\delta,\, l\beta}$$

Note that the matrix for the symmetry operation is orthogonal, that is,

$$\sum_{m\gamma} S^{-1}{}_{k\alpha,\, m\gamma} S^{-1}{}_{n\delta,\, m\gamma} = \delta_{kn} \delta_{\alpha\delta}$$

Then

$$\omega_l^2 S^{-1}{}_{n\delta,\, l\beta} = \sum_{k\alpha} \omega_k^2 S^{-1}{}_{k\alpha,\, l\beta} \delta_{kn} \delta_{\alpha\delta}$$

$$= \omega_n^2 S^{-1}{}_{n\delta,\, l\beta}$$

Since

$$\omega_l^2 \neq \omega_n^2$$

hence

$$S^{-1}{}_{n\delta,\, l\beta} = 0$$

This means that $\boldsymbol{S}$ is of the form:

$$\begin{bmatrix} \boldsymbol{\Gamma}^{(1)} & & & \\ & \boldsymbol{\Gamma}^{(2)} & & 0 \\ & & \boldsymbol{\Gamma}^{(3)} & \\ & 0 & & \ddots \end{bmatrix}$$

Each nonzero block matrix $\mathbf{\Gamma}^{(i)}$ corresponds to a mode. If Q_a is not degenerate, then

$$\mathbf{\Gamma}^{(a)} = \pm 1$$

If Q_a, Q_b are doubly degenerate, their $\mathbf{\Gamma}$ is

$$\begin{bmatrix} c_1 & c_2 \\ c_3 & c_4 \end{bmatrix}$$

c_1, c_2, c_3 and c_4 originate from the following relations:

$$SQ_a = c_1 Q_a + c_2 Q_b, SQ_b = c_3 Q_a + c_4 Q_b$$

Note that c_2 and c_3 can be nonzero. Of course for $S = E$, they are both zero.

The conclusion from the above analysis is that *normal coordinates are the bases of the irreducible representations of the molecular symmetry (point) groups*. Therefore, above $\mathbf{\Gamma}^{(i)}$s are irreducible. This is a very important property. It shows that normal coordinates can describe well the molecular symmetric properties and its complicated vibrations as shown in Chapter 3.

5.11 The reduction of the representation based on the atomic displacements

Suppose that ξ is the coordinate system of the atomic Δx, Δy, Δz displacements. Its relation with the normal coordinate is

$$\boldsymbol{Q} = \boldsymbol{M}\xi$$

Under the action of the symmetry operation $\boldsymbol{R}$:

$$\xi' = \boldsymbol{R}\xi$$

Note that

$$\begin{aligned} \boldsymbol{Q}' &= \boldsymbol{M}\xi' = \boldsymbol{MR}\xi = \boldsymbol{MRM}^{-1}\boldsymbol{M}\xi \\ &= \boldsymbol{MRM}^{-1}\boldsymbol{Q} \end{aligned}$$

In other words, representations by ξ and $\boldsymbol{Q}$ are equivalent. Under ξ, the representation matrix is $\boldsymbol{R}$ and under $\boldsymbol{Q}$, it is $\boldsymbol{MRM}^{-1}$ that is $\boldsymbol{S}$ in Section 5.10.

This analysis shows that as long as a coordinate system, which bears linear relation with the normal mode coordinates, is chosen, the symmetric properties of the normal modes can be easily retrieved by the group representation algorithm and even their explicit expressions are not known. Section 5.12 demonstrates this procedure.

5.12 The vibrational analysis of H_2O

For example, the point group of H_2O is C_{2v}. We choose the following atomic displacements as the basis:

$$\xi = \begin{bmatrix} \Delta x_{H_1} \\ \Delta y_{H_1} \\ \Delta z_{H_1} \\ \Delta x_{H_2} \\ \Delta y_{H_2} \\ \Delta z_{H_2} \\ \Delta x_O \\ \Delta y_O \\ \Delta z_O \end{bmatrix}$$

In calculating the character for an operation, it is not necessary to write down the explicit expressions of the whole matrix since what we need are only the diagonal terms that are 1, −1 or 0, depending on whether Δx_i is still Δx_i, $-\Delta x_i$ or changed to the others under the operation. By this rule, we have

$$\chi_\xi(E) = 9, \quad \chi_\xi(C_2) = -1$$
$$\chi_\xi(\sigma_v(zx)) = 1, \quad \chi_\xi(\sigma_v(yz)) = 3$$

This representation Γ_ξ can be reduced by eq. (5.3) or (5.4) (refer to the character table of C_{2v} in Appendix):

$$a_{A_1} = \frac{1}{4}[9 \cdot 1 + (-1) \cdot 1 + 1 \cdot 1 + 3 \cdot 1] = 3$$

$$a_{A_2} = \frac{1}{4}[9 \cdot 1 + (-1) \cdot 1 + 1 \cdot (-1) + 3(-1)] = 1$$

$$a_{B_1} = \frac{1}{4}[9 \cdot 1 + (-1) \cdot (-1) + 1 \cdot 1 + 3 \cdot (-1)] = 2$$

$$a_{B_2} = \frac{1}{4}[9 \cdot 1 + (-1) \cdot (-1) + 1 \cdot (-1) + 3 \cdot (1)] = 3$$

Or symbolically as

$$\Gamma_\xi = 3A_1 + A_2 + 2B_1 + 3B_2$$

From the character table, the translational and rotational movements are of B_1, B_2, A_1 (with x, y and z as the bases) and B_2, B_1, A_2 (with R_x, R_y and R_z as the bases) symmetries, respectively. Hence, the rest are the irreducible representations of vibration, that is,

$$\Gamma_{\text{vib}} = 2A_1 + B_2$$

The three normal modes are of symmetries, A_1, A_1 and B_2.

In turn, if the internal coordinates r_1, r_2 and α are employed (r_1 and r_2 are the two O–H stretches, α is the bend of HOH angle), their representations Γ_r and Γ_α can be reduced to

$$\Gamma_r = A_1 + B_2$$
$$\Gamma_\alpha = A_1$$

Hence, the mode of B_2 is pure stretch and the two modes of A_1 are composed of both the stretch and bending.

This example demonstrates that by the group representation theory, the symmetries of the normal modes, including their contents of internal coordinates, are very easy to obtain, though their detailed configurations (this is L_{ik}) are not known exactly. For this, the normal mode analysis is a necessity. The result is shown in Fig. 5.2.

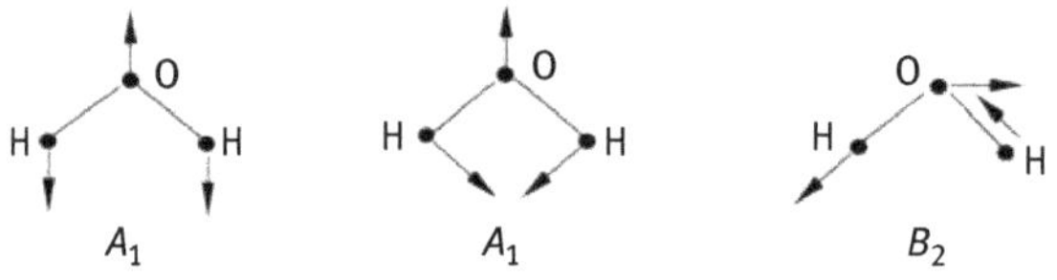

Fig. 5.2: The three normal mode configurations and their symmetries.

5.13 Wigner's projection operator

How to start with a variable υ (coordinate or function) to form $f^{(i)}(\upsilon)$ that is the basis of an irreducible representation $\Gamma^{(i)}$? Wigner's projection shows that $f^{(i)}(\upsilon)$ can be

$$f^{(i)}(\upsilon) = N\frac{d_i}{g}\sum_R \chi^{(i)}(R)\hat{R}\upsilon$$

where $\hat{R}\upsilon$ is the variable after υ is acted on by $\hat{R}$, d_i the dimension of the irreducible representation, g the group order, $\chi^{(i)}(R)$ the character of R in $\Gamma^{(i)}$ and N the normalization constant. $f^{(i)}(\upsilon)$ is also called the symmetric function of $\Gamma^{(i)}$.

5.14 Symmetry coordinates

From the internal coordinates, S_t's, the coordinate obtained by Wigner's projection operator is called the symmetry coordinate, S_i.

The point group of CO_3^{2-} is D_{3h}. The symmetries of its normal modes are

$$\Gamma_{vib} = A'_1 + A''_2 + 2E'$$

If the internal coordinates are chosen as shown in Fig. 5.3 (only the in-plane coordinates are considered):

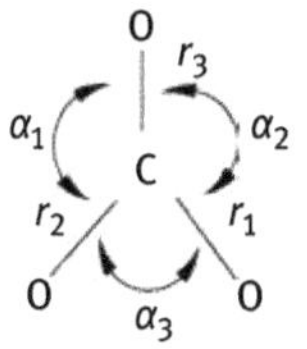

Fig. 5.3: The internal coordinates of CO_3^{2-}. Only the in-plane coordinates are considered.

By Wigner's projection operator, we have the symmetry coordinates:

$$S^r_{A'_1} = \frac{1}{\sqrt{3}}(r_1 + r_2 + r_3)$$

$$S^{ra}_{E'} = \frac{1}{\sqrt{6}}(2r_1 - r_2 - r_3)$$

$$S^{rb}_{E'} = \frac{1}{\sqrt{2}}(r_2 - r_3)$$

$$S^\alpha_{A'_1} = \frac{1}{\sqrt{3}}(\alpha_1 + \alpha_2 + \alpha_3)$$

$$S^{\alpha a}_{E'} = \frac{1}{\sqrt{6}}(2\alpha_1 - \alpha_2 - \alpha_3)$$

$$S^{\alpha b}_{E'} = \frac{1}{\sqrt{2}}(\alpha_2 - \alpha_3)$$

$$S^r_{A''_2} = 0 \; S^\alpha_{A''_2} = 0$$

$S^r_{A''_2}, S^\alpha_{A''_2}$ are zero since A''_2 corresponds to the out-of-plane motion. Note that A''_2 and z are of the same symmetry.

Since $\alpha_1 + \alpha_2 + \alpha_3 = 2\pi$, $S^\alpha_{A'_1}$ is equivalent to zero.

In the normal mode analysis, **F** and **G** are decomposed into 1×1, 2×2 and 2×2 matrices if $S^r_{A'_1}$, $\{S^{ra}_{E'}, S^{\alpha a}_{E'}\}$and $\left\{S^{rb}_{E'} S^{\alpha_b}_{E'}\right\}$ are adopted.

This example shows that if symmetry coordinates are adopted in the normal mode analysis, we are encountered with smaller matrices in the diagonalization of ***GF***. This is still helpful if the computing facilities are very common today since by this way, we can obtain the symmetry of the molecular vibration.

5.15 Direct product

For two groups $A=\{A_1, \ldots, A_f\}$, $B=\{B_1, \ldots, B_g\}$, it is shown that $\{A_1 \times B_1,\ A_1 \times B_2,\ \ldots,\ A_1 \times B_g,\ A_2 \times B_1,\ \ldots,\ A_f \times B_g\}$ forms a group. This is called the direct product group of A and B and is denoted by $A \times B$. The direct product group possesses the following properties:

1. Suppose that $[A_{ij}]_n$, $[B_{kl}]_m$ are the representation matrices for A and B groups, then $[A_{ij}B_{kl}]_{nm}$ is the representation matrix of the direct product $A \times B$. For example,

$$[A_{ij}]_n = \begin{bmatrix} A_{11} & A_{12} \\ A_{21} & A_{22} \end{bmatrix}, \quad [B_{kl}]_m = \begin{bmatrix} B_{11} & B_{12} \\ B_{21} & B_{22} \end{bmatrix}$$

   then

$$[A_{ij}B_{kl}]_{nm} = \begin{bmatrix} A_{11}B_{11} & A_{11}B_{12} & A_{12}B_{11} & A_{12}B_{12} \\ A_{11}B_{21} & A_{11}B_{22} & A_{12}B_{21} & A_{12}B_{22} \\ A_{21}B_{11} & A_{21}B_{12} & A_{22}B_{11} & A_{22}B_{12} \\ A_{21}B_{21} & A_{21}B_{22} & A_{22}B_{21} & A_{22}B_{22} \end{bmatrix}$$

2. The class number of the direct product group $A \times B$ is the product of the class numbers of A and B groups.
3. The number of irreducible representation of the direct product group $A \times B$ is the product of the numbers of irreducible representations of A and B groups.
4. The character of the representation matrix of the direct product group is the product of the characters of the representation matrices of A and B groups. This is due to the fact that

$$\begin{aligned} \chi_{nm} &= \sum_{il} [A_{ii}B_{ll}]_{nm} \\ &= \sum_{i} [A_{ii}]_n \sum_{l} [B_{ll}]_m \\ &= \chi_n \chi_m \end{aligned}$$

Hence, for the direct product groups, their character tables are easy to construct. For example, $D_{3h} = C_s \times D_3$, from the character tables (Tab.5.3 and Tab.5.4) of C_s and D_3, the character table of D_{3h} can be readily obtained as shown in Tab.5.5.

Tab. 5.3: Character table of C_s.

C_s	E	σ_h
A'	1	1
A''	1	−1

Tab. 5.4: Character table of D_3.

D_3	E	$2C_3$	$3C_2$
A_1	1	1	1
A_2	1	1	−1
E	2	−1	0

Tab. 5.5: Character table of D_{3h}.

D_{3h}	E	$2C_3$	$3C_2$	σ_h	$2S_3$	$3\sigma_v$
A'_1	1	1	1	1	1	1
A'_2	1	1	−1	1	1	−1
E'	2	−1	0	2	−1	0
A''_1	1	1	1	−1	−1	−1
A''_2	1	1	−1	−1	−1	1
E''	2	−1	0	−2	1	0

Now, we have the following remark:

We start with a set of functions $\{f_1, ..., f_n\}$ to form a representation. It can be reducible. Then, by similarity transformation, the representation matrices can be block-diagonalized to smaller irreducible representations. The bases corresponding to these irreducible representations are the linear combinations of $\{f_1, ..., f_n\}$, which can be obtained by the solvation of the similarity transformation.

The problem is: how is the similarity transformation obtained? How are $\{f_1, ..., f_n\}$ linearly combined to form the bases of the irreducible representations? This is not so obvious. In molecular vibration and electronic motion, the related issue is the solvation of the vibrational and electronic wave functions. This is, of course, not a straightforward problem.

From the character table and formulae like eqs. (5.3) and (5.4), we can obtain the irreducible representations in reducing a reducible representation. However, what we can have for their bases, most often, are only the functions that are of the same symmetries as the irreducible representations, that is, the symmetric functions. In other words, by group representation theory, we cannot obtain the eigenfunctions of the physical problems, say, the molecular vibration and electronic motion for which the solvation of equations of motion is required. In fact, the explicit expressions of the electronic wave functions or the normal coordinates in terms of the nuclear displacements are not so needy. Instead, their symmetries are more important and useful. Then, the group theoretical method is just powerful.

5.16 Symmetry of vibrational wave function

The wave function of the normal vibration in terms of normal coordinates is

$$\phi\{V_1V_2\cdots V_{3N-6}\} = \prod_{k=1}^{3N-6} \varphi_{V_k}(Q_k)$$
$$= \prod_{k=1}^{3N-6} \mathrm{N}_{V_k} e^{-\alpha_k Q_k^2/2} H_{V_k}(\sqrt{\alpha_k} Q_k)$$

where H_{V_k} is the Hermite polynomial, $\alpha_k = \omega_k/\hbar^2$

The wave function of the ground state is

$$\phi\{0, 0, \ldots, 0\} = N \exp\left[-\frac{1}{2}\sum_k \alpha_k Q_k^2\right]$$

If Q_k is not degenerate, under the symmetry operation

$$RQ_k = \pm Q'_k$$

then

$$Q_k^2 = Q'^2_k$$

For the degeneracy case, we have

$$R[Q_{k,1}, Q_{k,2}, \ldots, Q_{k,n}] = [Q'_{k,1}, Q'_{k,2}, \ldots Q'_{k,n}]$$

In matrix notation,

$$\mathbf{R}_k \mathbf{Q}_k = \mathbf{Q}'_k \tag{5.5}$$

Since the degenerate states possess identical ω_k and α_k, then

$$\begin{aligned}
&\alpha_k \sum_i Q_{k,i}^2 \\
&= \alpha_k \mathbf{Q}_k{}^T \mathbf{Q}_k \\
&= \alpha_k \mathbf{Q}_k{}^T \mathbf{R}_k^{-1} \mathbf{R}_k \mathbf{Q}_k \\
&= \alpha_k \mathbf{Q}_k{}^T \mathbf{R}_k{}^T \mathbf{R}_k \mathbf{Q}_k \\
&= \alpha_k \mathbf{Q}'_k{}^T \mathbf{Q}'_k \\
&= \alpha_k \sum_i Q'^2_{k,i}
\end{aligned} \tag{5.6}$$

Hence, the wave function of the ground state is of total symmetry (all characters are 1).

For the excited states, we have the following cases:

1. There is only one mode with quantum number 1, the others are 0:

$$\phi_{\left\{\begin{matrix} V_k=1 \\ V_{l\neq k}=0 \end{matrix}\right\}} = NQ_k \exp\left[-\frac{1}{2}\sum_l \alpha_l Q_l^2\right]$$

Since

$$\exp\left[-\frac{1}{2}\sum_l \alpha_l Q_l^2\right]$$

is of total symmetry, the symmetry of ϕ is that of Q_k.

2. There are two modes with quantum number 1, the others are 0:

$$\phi_{\left\{\begin{matrix} V_k=1 \\ V_l=1 \\ V_{m\neq k,l}=0 \end{matrix}\right\}} = NQ_kQ_l \exp\left[-\frac{1}{2}\sum_m \alpha_m Q_m^2\right]$$

The symmetry of ϕ depends on the symmetries of Q_k and Q_l. For example, for CO_3^{2-} (point group is D_{3h}), the symmetry of the wave function due to A'_1 and A''_2 modes

$$\phi_{\begin{bmatrix} V_{A'_1}=1 \\ V_{A''_2}=1 \\ V_m=0 \end{bmatrix}}$$

is

$$A'_1 \times A''_2 = A''_2$$

If the wave function is due to E' and E'', then

$$\phi_{\begin{bmatrix} V_{E'}=1 \\ V_{E''}=1 \\ V_m=0 \end{bmatrix}}$$

is of symmetry

$$E' \times E'' = A''_1 + A''_2 + E''$$

3. If the vibrational quantum number is larger than 2, for the nondegenerate case, the wave function is

$$\phi_{\left\{\begin{matrix} V_k = n_k \\ V_{l\neq k} = 0 \end{matrix}\right\}} = NH_{nk}(\sqrt{\alpha_k}Q_k)\exp\left[-\frac{1}{2}\sum_l \alpha_l Q_l^2\right]$$

As n_k is even, the Hermite polynomial is even. As n_k is odd, the Hermite polynomial is odd. Then as n_k is even, ϕ is of total symmetry. As n_k is odd, ϕ and Q_k belong to the same symmetry.

For the degeneracy case, the situation is more complicated. As an example, we consider the case of double degeneracy with $n_k = 3$. For this case, the distribution of quantum numbers is shown in Tab. 5.6.

We can choose Q_{ka}, Q_{kb} such that under the operation R,

Tab. 5.6: The distribution of quantum numbers for the case of double degeneracy with $n_k = 3$.

V_{ka}	V_{kb}	$\phi_{\left\{\begin{matrix} V_{ka}V_{kb} \\ V_l = 0 \end{matrix}\right\}}$
3	0	$H_3(Q_{ka}) \sim Q^3{}_{ka}$
2	1	$H_2(Q_{ka})H_1(Q_{kb}) \sim aQ_{ka}^2 Q_{kb} + bQ_{kb}$
1	2	$H_1(Q_{ka})H_2(Q_{kb}) \sim a_1 Q_{ka} Q_{kb}^2 + b_1 Q_{ka}$
0	3	$H_3(Q_{kb}) \sim Q_{kb}^3$

$$Q_{ka} \xrightarrow{R} R_a Q_{ka}$$
$$Q_{kb} \xrightarrow{R} R_b Q_{kb}$$

where R_a and R_b are constants.

Therefore, when

$$\begin{bmatrix} Q_{ka}^3 & \\ Q_{ka}^2 & Q_{kb} \\ Q_{ka} & Q_{kb}^2 \\ Q_{kb}^3 & \end{bmatrix}$$

is chosen as the basis, the character $\chi_3(R)$ for R is

$$\chi_3(R) = R_a^3 + R_a^2 R_b + R_a R_b^2 + R_b^3$$

Similarly for the case of $n_k = 2,\ n_k = 1,$ we have

$$\chi_2(R) = R_a^2 + R_a R_b + R_b^2$$
$$\chi_1(R) = R_a + R_b$$

Note that

$$Q_{ka} \xrightarrow{R^\upsilon} R_a^\upsilon Q_{ka}$$
$$Q_{kb} \xrightarrow{R^\upsilon} R_b^\upsilon Q_{kb}$$

and

$$\chi_1(R^\upsilon) = R_a^\upsilon + R_b^\upsilon$$

From $\chi_3(R)$, $\chi_2(R)$, $\chi_1(R)$, and $\chi_1(R^3)$, we have the expression:

$$\chi_3(R) = \frac{1}{2}\left[\chi_2(R)\chi_1(R) + \chi_1(R^3)\right]$$

This expression can be extended to

$$\chi_\upsilon(R) = \frac{1}{2}\left[\chi_{\upsilon-1}(R)\chi_1(R) + \chi_1(R^\upsilon)\right]$$

For the case of triple degeneracy, we have

$$\chi_\upsilon(R) = \frac{1}{3}(2\chi_1(R)\chi_{\upsilon-1}(R) + \frac{1}{2}\left\{\chi_1(R^2) - \chi_1^2(R)\right\} \cdot \chi_{\upsilon-2}(R) + \chi(R^\upsilon))$$

These expressions show that the character for the high quantum state can be conveniently obtained from those of low quantum states. As $\chi_\upsilon(R)$ is obtained, the representation can be reduced to reveal the symmetries of the vibrational states.

For example, for E' in D_{3h} with $n_k = 3$, the symmetries are $A'_1 + A'_2 + E'$. For the case with $n_k = 2$, the symmetries are $A'_1 + E'$.

4. The general cases

For a molecule with modes $v_1, v_2, \ldots, v_k, \ldots$ and the corresponding quantum numbers $V_1, V_2, \ldots, V_k, \ldots$ the symmetry of its wave function is $\Gamma = (\Gamma_1)^{V_1} \times (\Gamma_2)^{V_2} \times (\Gamma_3)^{V_3} \times \cdots \times (\Gamma_k)^{V_k} \times \cdots$ where Γ_k is the irreducible representation to which Q_k belongs.

For example, CO_3^{2-} has four modes: v_1, v_2, v_3, v_4 of A'_1, A''_2, E' and E', respectively. When $V_1 = 0, V_2 = 2, V_3 = 2, V_4 = 3$ the symmetries are:

$$\begin{aligned}\Gamma &= (A'_1)^0 \times (A''_2)^2 \times (E')^2 \times (E')^3 \\ &= A'_1 \times A'_1 \times (A'_1 + E') \times (A'_1 + A'_2 + E') \\ &= 2A'_1 + 2A'_2 + 4E'\end{aligned}$$

The key point is that $(E')^2$ (also for $(E')^3$) cannot be calculated simply from the character table by taking the squares of E' and then reducing it. The dimension obtained in this way is 4. In fact, the correct dimension is 3 as per the reason mentioned earlier. Of course, for two *different* modes of symmetry E', each with quantum number 1, the dimension is 4. Then, it is fine to take the square of E' and reduce it.

5.17 Selection rules

In Chapter 1, we know that the probability of transition from the initial state, $|I\rangle$ to the final state $|F\rangle$ by the process H' is

$$\langle F|H'|I\rangle$$

For this quantity to be nonzero, the product of their corresponding irreducible representations, $\Gamma_F\Gamma_{H'}\Gamma_I$, has to contain the totally symmetric irreducible representation for which all characters are 1.

We explain this rule by a case in which both $|I\rangle$ and $|H'\rangle$ possess functions f and h, respectively, and are of total symmetry. $|F\rangle$ has two functions g_1, g_2.

Hence, $\langle F|H'|I\rangle$ contains two situations:

$$\int g_1 hf d\tau \quad \text{and} \quad \int g_2 hf d\tau \tag{5.7}$$

Suppose that the representation by (g_1, g_2) can be reduced to two one-dimensional irreducible representations. Their bases G_1, G_2 are the linear combinations of g_1, g_2. Hence,

$$\int G_1 hf d\tau \text{ and } \int G_2 hf d\tau$$

are linear combinations of the two integrals in eq. (5.7). If G_1, G_2 are the bases of nontotally symmetric representations, there are operations O_1, O_2 such that

$$\int G_1 hf d\tau = O_1 \int G_1 hf d\tau = \int O_1 G_1 hf d\tau = -\int G_1 hf d\tau$$

$$\int G_2 hf d\tau = O_2 \int G_2 hf d\tau = \int O_2 G_2 hf d\tau = -\int G_2 hf d\tau$$

where the first term equals the second term since the integrals are constant and are invariant to O_1, O_2. Hence, $\int G_1 hf d\tau$ and $\int G_2 hf d\tau$ are zero. So are $\int g_1 hf d\tau$ and $\int g_2 hf d\tau$.

If G_1 (or G_2) is the basis of the totally symmetric representation, then the integral can be nonzero. But, we cannot assure that it is definitely nonzero. In other words, the transitional probability is possible. However, if $\Gamma_F\Gamma_{H'}\Gamma_I$ does not contain the totally symmetric representation, then the corresponding integral will definitely be zero, that is, the transitional probability is zero.

From this analysis, we have in general that:

$$\int f_1 f_2 ... d\tau \propto \int g d\tau$$

where g is the linear combinations of $f_1, f_2, ...$ and is of the totally symmetric irreducible representation.

Now back to the original issue. For the dipolar process, the transition matrix is $<F|\mu|I>$ with μ the dipole operator. The selection rule is that $\Gamma^F \times \Gamma^\mu \times \Gamma^I$ contains the totally symmetric representation.

For example, Q_{v_1} in CO_3^{2-} belongs to A'_1, and Q_{v_3} belong to E'. For the v_1 mode, the transition from $n_{v_1} = 0$ to 1 is not i.r. (infrared) active since

$$\begin{aligned} &\Gamma^F \times \Gamma^\mu \times \Gamma^I \\ &= A'_1 \times \begin{pmatrix} E'(x,y) \\ A''_2(z) \end{pmatrix} \times A'_1 \\ &= \begin{pmatrix} E' \\ A''_2 \end{pmatrix} \end{aligned}$$

does not contain A'_1.

The transition from $n_{v_1} = 0$, $n_{v_3} = 1$ to $n_{v_1} = 1$, $n_{v_3} = 1$ is i.r. active with x, y polarizations, while the z polarization is not allowed. This is because that

$$\begin{aligned} &\Gamma^F \times \Gamma^\mu \times \Gamma^I \\ &= (A'_1 \times E') \begin{pmatrix} E'(x,y) \\ A''_2(z) \end{pmatrix} \times E' \\ &= \begin{pmatrix} 3E' + A'_1 + A'_2 \\ A''_1 + A''_2 + E'' \end{pmatrix} \end{aligned}$$

5.18 Correlation

When a molecule changes its configuration, its symmetry group changes, so do the symmetries of its wave functions. The change of symmetry will lead to different selection rules. However, the wave functions may not change much, especially when the configuration change is a small perturbation. In other words, there is correspondence between the irreducible representations of the two symmetry groups before and after the change. This correspondence is called correlation.

For instance, A' and A'' of C_s and A_g, A_u, B_g, B_u of C_{2h} are correlated as shown in Fig. 5.4.

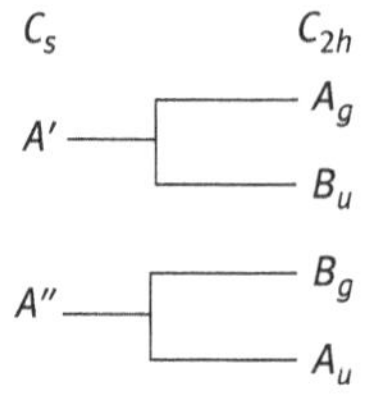

Fig. 5.4: The correlation of A' and A'' of C_s and A_g, A_u, B_g, B_u of C_{2h}.

In the free space, ClO_3^- is of C_{3v} group. Its four modes, v_1, v_2 and v_3, v_4 belong to A_1 and E symmetries, respectively. As ClO_3^- is in the $KClO_3$ crystal, due to the environmental change, the potential field around ClO_3^- is no longer C_{3v}. Instead, it is C_s. The symmetries of its vibrational modes also change. The correlations of the irreducible representations among these two groups are shown in Fig. 5.5.

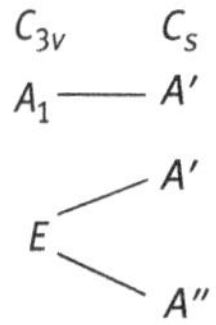

Fig. 5.5: The correlation between C_{3v} and C_s.

This correlation shows that the degenerate v_3, v_4 modes, which are of E symmetry in C_{3v}, decompose into two nondegenerate modes of A' and A'' symmetries, respectively, in $KClO_3$ crystal.

It happens often that the understanding of the detailed vibrational configurations is not so necessary. Instead, it is their symmetries that are more important and useful. In this section, we demonstrate that simply by correlation, the splitting of modes is easily deduced when the potential field around a molecule is changed.

For correlation among irreducible representations of various groups, see References [1, 2].

5.19 Comments on the point groups

1. For the cyclic group C_n, its each element forms a class. It has n classes and n irreducible representations. By eq. (5.2), we have

$$\sum_{i=1}^{n} d_i^2 = n$$

Hence, $d_i = 1$. This means that its irreducible representations are of one dimension.

However, in the character table of C_n, there are irreducible representations labeled by E. Is there any contradiction?

The correct understanding is: In general, there is no degeneracy among the different one-dimensional irreducible representations. However, for C_n, there is a situation that its two irreducible representations are conjugate to each other. Their characters are complex and complex conjugate to each other. Suppose that ψ_i is the basis for the one irreducible representation, then ψ_i^* is the basis for the other conjugate irreducible representation. If ψ_i is the eigenstate of the C_n system, that is:

$$H\psi_i = E_i\psi_i$$

then

$$H\psi_i^* = E_i\psi_i^*$$

This means that ψ_i and ψ_i^* are degenerate.

If a real variable is chosen for the bases of the two irreducible representations, there is another real variable such that these two variables together form the bases of the two mutually conjugate irreducible representations. The representation by these two real variables is two dimensional and reducible, which can be reduced to the two irreducible representations.

This explains that for the C_n character table, there are conjugate one-dimensional irreducible representations labeled as E, E_1, E_2 and their bases are two dimensional (x, y), (R_x, R_y) and so on.

2. Two-dimensional irreducible representations can possess basis variables such as (x, y), (R_x, R_y). However, for C_1, C_s, C_2, C_i, C_{2h}, there are one-dimensional irreducible representations with bases such as (x, y), (R_x, R_y). Obviously, the variables (x, y), (R_x, R_y) are not two dimensional. They are one-dimensional variables (e.g., dipole) lying somewhere along (x, y), (R_x, R_y) coordinates.

5.20 Comments on the quantum numbers

In solving the Schrödinger equation, we demand that the wave functions are zero on the boundary. This leads to quantization that is associated with the quantum numbers. In different coordinate systems (Cartesian or spherical), the forms of quantum numbers are different.

In fact, quantum number is closely related to the classical action. In the Hamilton–Jacobi mechanics, action and angle form a pair of dynamical variables. Quantum number is the characterization of conserved action. The essence of quantum number is the constant of motion. We note that the constant of motion is the result of symmetry. Hence, quantum number is related to the symmetry group. We will show this relation by the discrete translational group. It will be demonstrated that quantum number can be the labeling for the irreducible representation.

The irreducible representations of a group are the realization of its classes. An important property is that the number of irreducible representations and the class number are the same. This hints that irreducible representation and class are of the reciprocal relation.

For the translational group, since its elements commute, each class is composed of only one element. For a system of N unit cells, there are N elements and N classes: $\boldsymbol{T}_0(=\boldsymbol{E})$ (identity), $\boldsymbol{T}_1, \ldots, \boldsymbol{T}_{N-1}$ and N irreducible representations. The characters are of the form $\exp(i\boldsymbol{k}_j\boldsymbol{T}_n)$ with $\boldsymbol{k}_j$ the wave vector. We can employ $\boldsymbol{k}_0$ (totally symmetric representation), $\boldsymbol{k}_1, \ldots, \boldsymbol{k}_{N-1}$ to label the irreducible representations. These wave vectors are quantized. $\boldsymbol{T}_n$ is the coordinate and $\boldsymbol{k}_j$ is the quantization of action ($\hbar\boldsymbol{k}_j$). They are the action–angle variables. This means that quantum numbers are the labels for the irreducible representations. The relation of irreducible representation and class is just like that of action and angle. This is a universal result.

For a quantal system, the employment of different coordinate systems (or symmetry groups) will result in different forms of quantum numbers. These different forms of quantum numbers describe the identical quantum system. The difference lies in the different physical perspectives. They help the different classifications of its quantum states and the revelation of their properties.

Exercises

5.1 Show that S_n (n odd) axis implies the existence of C_n axis and σ_h.

5.2 Show that the finite cyclic groups of the same order are isomorphic.

5.3 Show that the group of order 5 is a cyclic group.

5.4 Show that each element forms a class by itself for the abelian groups.

5.5 For the subgroup $\{E, C_3, C_3^2\}$ of C_{3v}, write down the right and left cosets and the classes of C_{3v}.

5.6 Show that two cosets are identical if they share a common element.

5.7 Show in Section 5.3.3 that

$$FK_i = K_i,\ K_iK_i^{-1} = F$$
$$(K_iK_j)K_k = K_i(K_jK_k)$$

5.8 Show that a group is homomorphic to its quotient group.

5.9 Show the groups of $C_3 \times C_i$, $C_{3v} \times C_i$, $C_{5v} \times C_s$, $T_d \times C_i$

5.10 Show the point groups of molecules in Fig. 5.6

5.11 Show that the characters of matrices of the equivalent representations are the same.

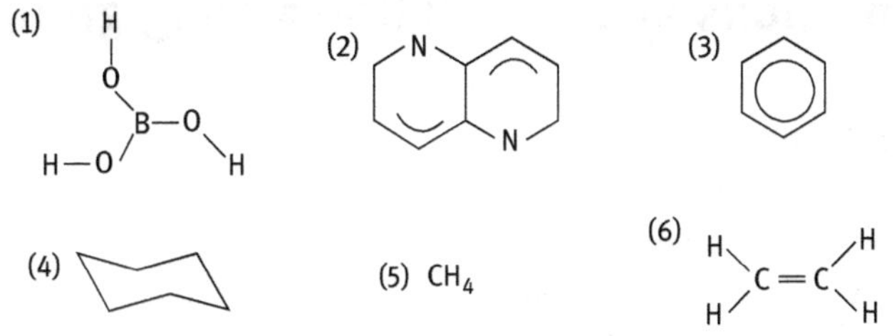

Fig.5.6: The molecular structures for Exercise 5.11.

5.12 Show in Section 5.7 that

$$\sum_i d_i^2 = g$$

5.13 For the following system as shown in Fig.5.7

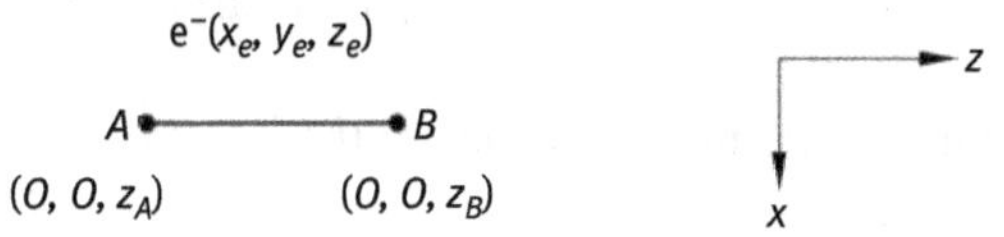

Fig. 5.7: The system for Exercise 5.13.

show that the potential is invariant under the rotation of $\pi/2$ and π along the x-axis. A and B are nuclei, and e^- is an electron.

5.14 Show the normal modes of

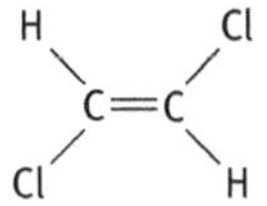

Which are i.r. or Raman active (see Chapter 8)? Which are in-plane and out-of-plane motion? If the symmetry is reduced to C_2, how the symmetries of the normal modes will change?

5.15 Write down the symmetry coordinates of the above system.

5.16 Express the potential matrix $\boldsymbol{F}$ by the internal and symmetry coordinates of the following molecule:

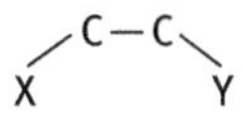

References

[1] Cotton FA. Chemical Applications of Group Theory 3rd.ed. New York: Wiley, 1990.
[2] Wilson EB, Decius JC, Cross PC. Molecular Vibrations. New York: McGraw Hill, 1955.

6 Crystal vibrations and correlation among groups

6.1 Crystal vibrations

One of the characteristics of molecular crystals is their periodic arrays. The whole crystal structure is constructed from a small *unit cell* by the translational symmetry. In a unit cell, there are multiple (or one) molecules (including cations and anions) and symmetry elements among its various sites that form the unit cell group. The translational symmetries form the translational group. The product of the unit cell group and translational group forms the space group, which describes the symmetric properties of the crystal as a whole.

We view the whole crystal as a large molecule. Suppose that a crystal contains N unit cells, each unit cell contains P molecules and each molecule contains n atoms. Then, the whole number of normal modes is $3nPN$. Since N is of the order of 10^{23}, the number of normal modes is 10^{23}. There is very regular distribution of these modes as will be shown below.

First, we note that for each molecular normal mode, all the bond stretches and bends possess the same vibrating frequency, and the phase relations among them are fixed at 0 or π. In a crystal, the normal modes among the various unit cells should also possess fixed phase relations. Since the intermolecular interaction in a crystal is much smaller than the intramolecular interaction among the atoms in a molecule, these phases may not be 0 or π. The phase relation between the same normal modes of equivalent molecules in any two unit cells can be any number in between 0 and π. The corresponding wavelength λ is from ∞ to $2d$ (where d is the dimension of the unit cell), or the range of the wave vector k (defined as $2\pi/\lambda$) is

$$0 \le k \le \pi/d$$

Wave vector is directional. In a one-dimensional case, we have

$$-\pi/d \le k \le \pi/d$$

For the three-dimensional case, the wave vector is denoted as $\boldsymbol{k}$. Besides, we have to note that the medium in a crystal is not continuous. Hence, the wavelength and wave vector cannot be continuous and their total numbers are finite. It can be shown that the number of allowed $\boldsymbol{k}$ is exactly N.

Here, we only mentioned the phase relation between the same normal modes of the equivalent molecules in different unit cells. This is a consequence of translational symmetry.

Up to this point, we can understand that for each molecular normal mode, there are N crystal modes derived from it. These crystal modes can be labeled by $\boldsymbol{k}$. Since there is long-range interaction among the equivalent molecules in different unit cells, such as the dipolar coupling among the vibrationally induced dipole

https://doi.org/10.1515/9783110625097-006

moments, the frequencies of these crystal modes will be dispersive on $\boldsymbol{k}$. This is called dispersion. Since the interaction is not large, the dispersion cannot be serious.

The aforementioned analysis shows that, to each molecular normal mode, there is a branch distribution of crystal modes as shown in Fig. 6.1. Obviously, there are $3nP$ branches.

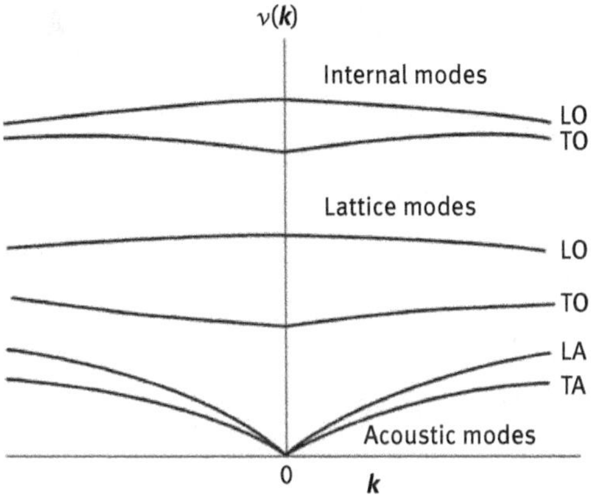

Fig. 6.1: The dispersion relation of crystal modes. L, T, O and A denote the longitudinal, transverse, optic and acoustic modes, respectively.

These branches of crystal modes can be classified as follows:

1. The internal modes originate from the molecular normal modes.
2. The external modes originate from the relative motions of cations and anions like the lattice modes.

These two modes are called the optic modes since they can couple with light. The coupling is due to the vibrationally induced dipole moment and the electric field of light.

Besides, there is one more mode:

3. The acoustic modes originate from the motions of cations and anions that are in the same direction. These modes cannot produce vibrationally induced dipole moments and, hence, cannot couple with light.

The difference between the acoustic and optic modes is that as $\boldsymbol{k} \to 0$, the frequency of the acoustic mode approaches 0, while that of the optic mode approaches a non-zero value. Fig. 6.2 shows their characters.

We note that each crystal mode can be labeled by $\boldsymbol{k}$. If the movement of the particles or the direction of the induced dipole moments of a crystal mode is parallel to $\boldsymbol{k}$, the mode is called the longitudinal mode. If it is vertical to $\boldsymbol{k}$, the mode is called the transverse mode. The transverse mode is degenerate (the plane vertical to $\boldsymbol{k}$ is two-dimensional) and its frequency is less than the longitudinal mode.

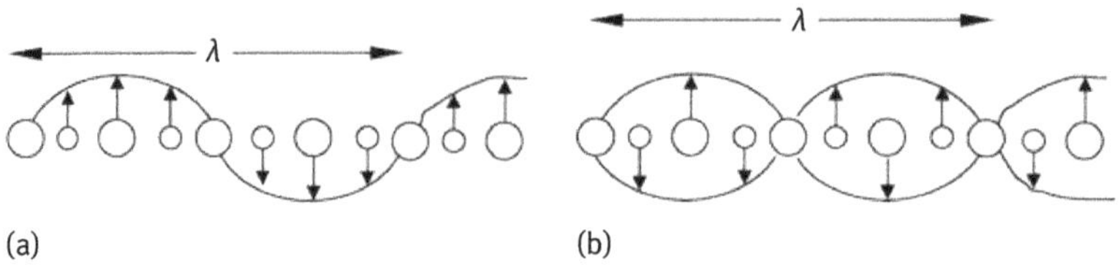

Fig. 6.2: Acoustic mode (a) and optic mode (b). Circles represent the cations and anions. Arrows show the displacements.

Crystal modes can be quantized. The quantum is called phonon.

The selection rules for the coupling between phonon and photon are as follows:

1. Energy conservation:

$$\hbar\omega = \hbar\omega_q$$

   where ω and ω_q are the angular frequencies of photon and phonon.

2. Momentum conservation:

$$\hbar\boldsymbol{k} = \hbar\boldsymbol{q}$$

   where $\boldsymbol{k}$ and $\boldsymbol{q}$ are the wave vectors of photon and phonon.

   Since the wavelength of (visible) light is much larger than the unit cell dimension, only the crystal modes with $\boldsymbol{k}$ close to 0 can couple with light.

3. Other symmetry requirements: This will be discussed latter.

It is obvious that longitudinal mode cannot couple with light.

These rules, especially the momentum conservation, lead to an inference that only a very limited number of crystal modes $\approx 3nP$ among the $3nPN \approx 10^{23}$ modes can be optically active. In other words, the i.r. and Raman spectra of molecular crystals are very similar to those of gas, liquid and solution. The differences are some external modes, which do not appear in gas, liquid and solution, and the splitting of the degenerate internal modes. Of course, because of couplings, the mode frequencies may shift slightly. All these can be explained by the considerations based on symmetry. For a quantitative analysis, details about the couplings in the crystals are needed. This is the field of crystal dynamics.

6.2 Unit cell group, site group and translational group

In a unit cell, there are symmetry operators relating its various sites. This is called the unit cell group. The unit cell group is different from the point group in that there are no fixed points. However, a unit cell group can be isomorphic to a point group. This enables us to discuss the crystal vibrations based on the point groups.

Consider $LiKSO_4$ crystal as an example. The unit cell is hexagonal with dimensions: $a_0 = b_0 = 5.1457\ \mathring{A}$, angle 120° and c_0=8.6298 $\mathring{A}$. The symmetry operations in the unit cell are shown in Fig. 6.3. It includes the following:

1. C_3 operation, denoted by ▲.
2. C_2 operation, then sliding along c-axis with a displacement of $c_0/2$, denoted by $(C_2, c/2)$.
3. C_6 operation, then sliding along c-axis with a displacement of $c_0/2$, denoted by $(C_6, c/2)$.

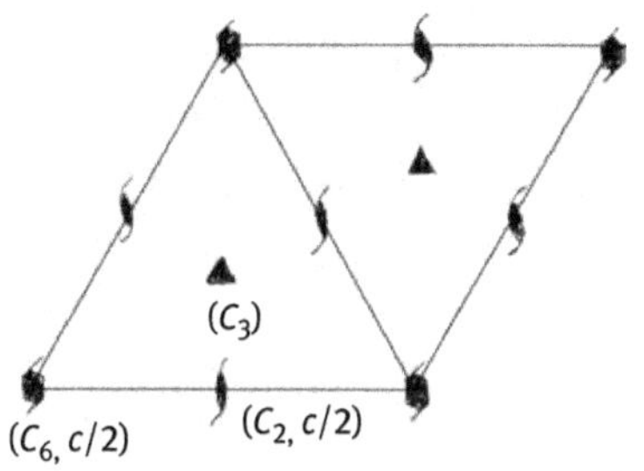

Fig. 6.3: The unit cell group operations of $LiKSO_4$ crystal.

The unit cell group operations are E, $(C_2, c/2)$, C_3, $(C_6, c/2)$, C_3^2, $(C_6^5, c/2)$. They satisfy the definition of a group. For these operations, we have to note the translational equivalence such as

$$(C_6, c/2) \bullet (C_6, c/2) = (C_6^2, c) = (C_3)$$

The operations of a unit cell group are different from those of its isomorphic point group in that they have their own rotation axes and, as mentioned earlier, sliding along the c-axis with $c_0/2$ and other characters that are not found in the point groups. These details are not discussed here; they are described in the books on the X-ray structure determination.

The isomorphism between the unit cell and point groups can be identified as shown in Tab. 6.1:

Tab. 6.1: The isomorphism between the unit cell and point groups.

Unit cell group		Point group
E	→	E
$(C_6, c/2)$	→	C_6
C_3	→	C_3
$(C_2, c/2)$	→	C_2
C_3^2	→	C_3^2
$(C_6^5, c/2)$	→	C_6^5

Because of the isomorphism, we can use the point group notation for the unit cell group. For $LiKSO_4$, its unit cell group can be designated by C_6.

Previously, we mentioned that space group is a product of a unit cell group and the translational group. Hence, we may have different space groups that possess the same isomorphic unit cell groups. For these space groups, we can add superscripts, which are by the international convention, to the isomorphic point group to make them distinguishable. For instance, the space group of $LiKSO_4$ is C_6^6.

Hence, there are many space groups. However, the number of unit cell groups is very limited. In fact, there are 230 space groups and only 32 unit cell groups.

The unit cell group is sometimes called the quotient group. This is because of the fact that the translational group is the invariant subgroup of the space group and its corresponding quotient group is the unit cell group. Conceptually, we can consider the unit cell group as the *quotient* (a group) of the space group by the translational group.

The order of the translational group is the number of unit cells, N. Translational group is commutative (abelian) and its irreducible representations are all one dimensional (see Section 5.7). Hence, there are N one-dimensional irreducible representations. In the last section, it was mentioned that the number of wave vectors $\boldsymbol{k}$ is also N. This is not accidental. In fact, $\boldsymbol{k}$ can be employed to label the irreducible representations of the translational group. Their characters are:

$$e^{i\boldsymbol{k}\bullet\boldsymbol{R}_n}$$

$\boldsymbol{R}_n$ is the position vector of a unit cell with respect to an arbitrarily chosen origin. Different $\boldsymbol{k}$ shows different irreducible representations and different $\boldsymbol{R}_n$ shows different translational elements.

The positions of Li, K, S and O in a unit cell of $LiKSO_4$ crystal are designated as x, y and z, which are less than 1 and greater than 0. They are referred to the cell dimensions along the a, b and c axes. For a point with x, y or z larger than 1 or less than 0, it can be translated back to the previously designated equivalent point by translational symmetry.

In each unit cell of $LiKSO_4$ crystal, there are two molecules. Li, K and S atoms are at

K (2*a*); (0 0 0); (0 0 1/2)
S (2*b*); (1/3, 2/3, 0.17); (2/3, 1/3, 0.67)
Li (2*b*); (1/3, 2/3, ~1/2); (2/3, 1/3, ~0)

Here, the positions of O atoms are not shown. From the viewpoint of crystal dynamics, the position of S atom and the potential around it can be employed for the dynamical analysis of SO_4 group.

In the above notations, 2 is the number of atoms and *a, b* denote the site symmetries. Site symmetries are the symmetries of the potential around a site. The symmetries form a site group. For each unit cell group, the site groups that the notations *a, b* represent can be found from Reference [1]. (In fact, the symmetries at a site are easy to be figured out.) Site groups are the point groups in which the fixed points are the sites themselves.

6.3 Molecular point group, site group, unit cell group and their correlation

Molecular point group is the geometric symmetry of a molecule in free space. When a molecule (or the species like the cation or anion) is introduced into the crystal, the symmetries of the potential around the site where the molecule sits form a site group. The normal modes of the molecule in free space can be classified by the irreducible representations of the molecular point group. As it is at the site in the crystal, the symmetries and the degeneracies of its modes will change. Then, its modes can be classified by the irreducible representations of the site group. There is correlation between these two sets of classifications. Since the crystal field is much smaller than the bond force field in the molecule, the mode configuration remains almost unchanged. The physical background of the correlation from the molecular point group to the site group is the effect of the crystal field on the molecular vibrational modes.

In a unit cell, the molecular (including the ions) vibrational modes at various sites are not absolutely irrelevant due to the fact that there are intermolecular couplings and that all the modes have to be the bases of the irreducible representations of the unit cell group. Mathematically, by Wigner's projection operator, we can construct the bases of the irreducible representations of the unit cell group as the linear combinations of the *variables* (mode, dipole, coordinate) at the various sites in the unit cell. The variables at the various sites in the unit cell are related to each other by the symmetry operators of the unit cell group. If the variable is the vibrational mode, then the picture is that the unit cell mode is the linear combinations of the molecular normal modes at the various sites in the unit cell that are related by the symmetry operators of the unit cell group. The correlation of the site group to the unit cell group shows how the bases (molecular vibrational modes) of the irreducible representations of the site group are linearly combined to form the basis (unit cell mode) of the irreducible representation of the unit cell group.

It is then an analogy: crystal modes are derived from the unit cell modes by Wigner's projection operator based on the translational symmetries. For instance, if Q_n is the coordinate of a unit cell mode in the *n*th unit cell, the corresponding crystal mode is:

$$\sum_n Q_n e^{i\boldsymbol{k}\bullet\boldsymbol{R}_n}$$

where $\boldsymbol{R}_n$ is the position vector of the nth unit cell.

Since only those crystal modes with $\boldsymbol{k} \approx 0$ can couple with light, the crystal mode coordinate then reduces to Q_n. Hence, we only need the unit cell mode coordinates.

Dynamically, for $\boldsymbol{k} \neq 0$, the unit cell group cannot faithfully show the crystal vibrational dynamics. Then, the dynamical quantities in different unit cells are not the same. There are phase differences among them. The dynamics with $\boldsymbol{k} \neq 0$ is crucial for the phase transition. During the phase transition, the unit cell structures may vary: some are of the low-temperature phase structure and some are of the high-temperature phase structure.

The correlation among the molecular point group, site group and unit cell group can be found in many books [2].

In the following, the case of $LiKSO_4$ crystal will be demonstrated [3,4].

SO_4 has four modes, ν_1, ν_2, ν_3 and ν_4. Under T_d, they belong to A_1, E, T_2, T_2 symmetries, where E and T_2 are two and three dimensional, respectively.

In the crystal, Li, K, S are at a or b sites. Reference [1] shows that both site groups are C_3. The unit cell group is C_6. The correlation among these groups is shown in Fig. 6.4.

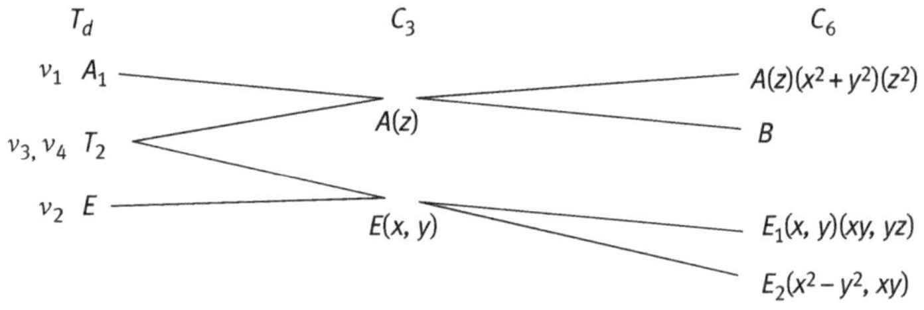

Fig. 6.4: The correlation among T_d, C_3 and C_6.

Now, we show how the variables at the sites where the ions sit are linearly combined to form the bases of the irreducible representations of the unit cell group. Let Li(1), Li(2), K(1), K(2), S(1), S(2) denote the variables at the sites where Li, K, S sit. By Wigner's projection operator, the bases of A, B, E_1 and E_2 irreducible representations of C_6 are

$$
\begin{aligned}
A,\ E_1:\ & 1/\sqrt{2}[Li(1)+Li(2)] \\
& 1/\sqrt{2}[K(1)+K(2)] \\
& 1/\sqrt{2}[S(1)+S(2)]
\end{aligned}
$$

$$
\begin{aligned}
B,\ E_2:\ & 1/\sqrt{2}[Li(1)-Li(2)] \\
& 1/\sqrt{2}[K(1)-K(2)] \\
& 1/\sqrt{2}[S(1)-S(2)]
\end{aligned}
$$

Their physical significances are shown below. From the correlation, it is known that under the site group C_3, the triply degenerate ν_3 and ν_4 modes are split into one- and two-dimensional modes. In the unit cell, the modes of the two SO_4 ions are then combined with phase 0 or π to form A, E_1 and B, E_2 modes, where A, E_1 are of z and (x, y) symmetries, respectively, and are i.r. active. If we treat S(1), S(2) as the vibrationally induced dipole moments, then the linear combinations can be depicted as shown in Fig. 6.5. In the figure, A, E_1 modes possess net dipoles in a unit cell while for B, E_2 modes, there are no net dipoles though at sites (1), (2), there are the dipoles. Since the wavelength of i.r. is much longer than the unit cell dimension, i.r. can only couple with A, E_1 modes. Light that can couple with the B, E_2 modes will have wavelength in the order of the unit cell dimension. Then, it is in the X-ray range which is of too high energy for the vibrational coupling to occur.

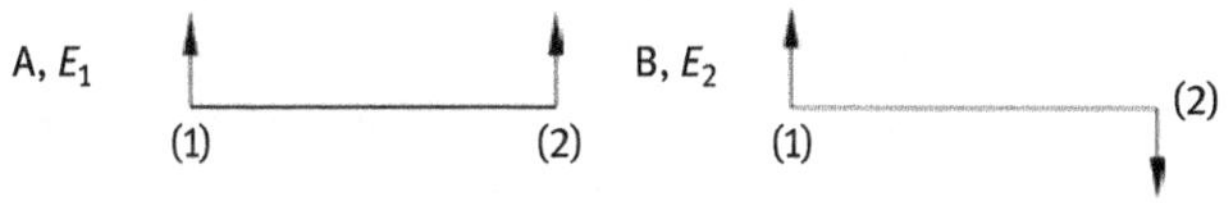

Fig. 6.5: The geometric representations of A, E_1, B and E_2 of the unit cell group. (1) and (2) are the two sites that are related by the symmetries of the unit cell group.

In the following, the lattice modes will be addressed.

The translations of SO_4 along the crystal axes z and x, y are of the symmetries A and E, respectively. As shown earlier, these SO_4 translations can be combined to form the A, B, E_1, E_2 crystal lattice modes. The same procedure can be applied for the Li, K crystal lattice modes. Strictly speaking, the lattice modes of Li, K, SO_4 and the crystal modes originating from the SO_4 vibrational modes will mix somewhat as long as they are of the same symmetry due to the higher order couplings. The effect is mostly reflected in the Raman or i.r. intensities.

From this example, we understand that the order (the number of the symmetry operations) of the unit cell group is just an integral multiple, P (the number of molecules in a unit cell), of the order of the site group, that is, site group is a subgroup of the unit cell group. In correlation, the irreducible representation of d dimensions of the site group correlates exactly to the irreducible representations of the unit cell group of total dimensions $P \cdot d$. This property is very convenient for checking correlation.

Another example is the case of K_2SO_4 [3,4].

The space group of K_2SO_4 crystal is D_{2h}^{16}. The site groups for K and S atoms are C_S. The correlation for the point group, site group and the unit cell group for SO_4 is shown in Fig. 6.6.

The irreducible representations A', A″ of C_S are of (x, y) and z symmetries. Since A' is one dimensional, its variable (x, y) is also of one dimension. If (x, y) is considered as a vector (dipole), its direction points somewhere in the x–y plane with both x- and y-components.

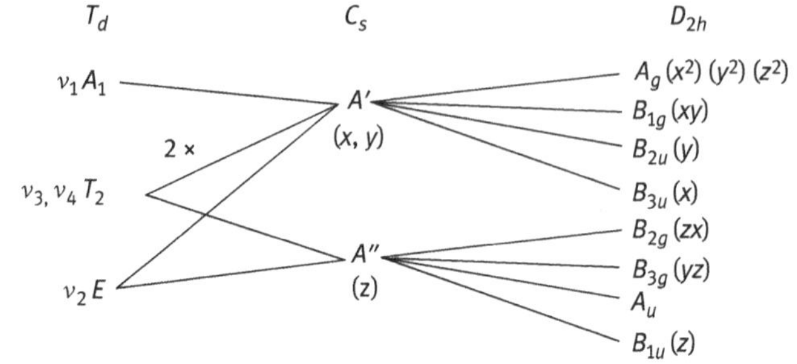

Fig. 6.6: The correlation among T_d, C_S and D_{2h}.

Figure 6.7 shows the components of the irreducible representations of the unit cell group where 1, 2, 3 and 4 denote the four sites where SO_4 sits. A_g, B_{1g}, B_{2g}, B_{3g}, A_u modes possess no net dipoles though there are dipoles at individual sites. Hence, they cannot be i.r. active. These modes besides A_u are Raman active. B_{1u} mode possesses the z-directional dipole and is i.r. active. B_{2u} and B_{3u} modes possess both x- and y-components at the individual sites. However, because of cancellation, they only possess net y and x components, respectively.

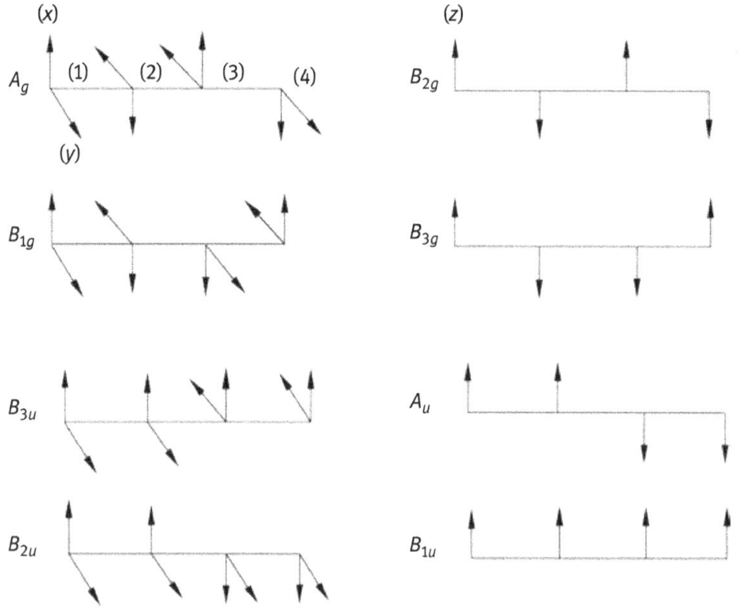

Fig. 6.7: The geometric representations of the irreducible representations of D_{2h} for K_2SO_4. (1), (2), (3) and (4) are the four sites where SO_4 sits. They are related by the symmetries of the unit cell group.

From the D_{2h} character table, it is seen that B_{2u} and B_{3u} are, respectively, of y and x symmetries. This is a global property. From the structure of the unit cell, both B_{2u} and B_{3u} modes possess implicit x- and y-components, respectively. This is due to the one dimensional A' of site group C_s, which possesses both the x- and y-components. For B_{1u} mode, A'' is one dimensional with z-component only. Hence, it only possesses dipoles along the z-direction.

References

[1] Decius JC, Hexter RM. Molecular Vibrations in Crystals. New York: McGraw-Hill, 1977.
[2] Wilson EB, Decius JC, Cross PC. Molecular Vibrations. New York: McGraw-Hill, 1955.
[3] Wu G, Frech R. J. Chem. Phys. 1976, 64: 4897.
[4] Wu G, Frech R. J. Chem. Phys. 1977, 66: 1352.

7 Electronic wave functions

7.1 Electronic wave functions

In Chapter 8, we will discuss about Raman effect. This effect is a result of coupling of the electronic and vibrational motions (called vibronic coupling). Hence, in this chapter, we introduce the concept of the electronic motion.

The electronic motion in a molecule is described by the wave function. This description is not completely the same as that of the nuclear motions – rotation and vibration. This is because the motion speed of electrons is much faster (see Section 2.2). The theory that deals with the electronic motion is called the molecular orbital (MO) theory, where the concept of orbital is employed. The MO theory is often called quantum chemistry since it is commonly used in chemistry.

7.2 Linear combination of atomic orbitals

The electrons in a molecule move in a very complex potential field. Besides the attraction between the electrons and nuclei, there is the repulsion among the electrons. Therefore, the rigorous solution for the electronic motion is very difficult to obtain. Approximation cannot be avoided. The algorithm based on the linear combination of atomic orbitals (LCAO) is an effective and convenient one.

Before the introduction of LCAO, we recall the interference on the screen when two light beams that are emitted from a point source and after passing through two pinholes finally reach the screen as shown in Fig. 7.1. Because of the phase difference between the two light beams, there are interference (bright and dark) lines on the screen. In Fig. 7.1, ϕ_1 and ϕ_2 denote the states (or wave functions) of the two beams after they pass through the pinhole. The bright and dark lines on the screen are $|\phi_1 + \phi_2|^2$ because of the combination of ϕ_1 and ϕ_2. Note that

$$\begin{aligned}|\phi_1 + \phi_2|^2 &= (\phi_1 + \phi_2)(\phi_1^* + \phi_2^*)\\ &= |\phi_1|^2 + |\phi_2|^2 + \phi_1\phi_2^* + \phi_1^*\phi_2\end{aligned}$$

The line intensity is a result of not only the addition of the original two beams, $|\phi_1|^2 + |\phi_2|^2$, but also their interference $\phi_1\phi_2^* + \phi_1^*\phi_2$. Interference may enhance or suppress the intensity, resulting in bright and dark lines.

Now, we know that interference or resonance is due to linear combination of states. The origin is the *indistinguishability* of the two pinholes as far as the light beam is concerned. In Chapter 5, we have discussed in detail the ideas of symmetry and group. What is symmetry? It is indistinguishability.

https://doi.org/10.1515/9783110625097-007

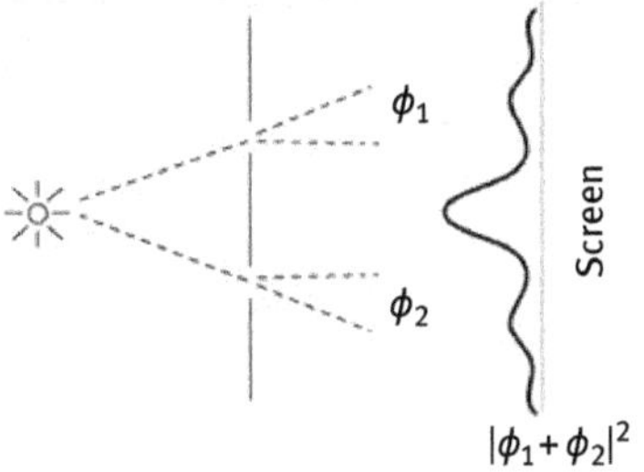

Fig. 7.1: The interference by the light beams after passing through the pinholes.

In conclusion, *linear combination, indistinguishability* (*symmetry*) and *resonance* are closely related ideas.

With these ideas, we can now explore the electronic wave functions. (1) Although there are many electrons in a molecule, as a first approximation, we may consider that each electron moves with a potential resulting from the complicated interactions among electrons and nuclei. This is the so-called one-electron model. (2) As a molecule is formed, the electrons that were originally in the individual atoms can now migrate among the atoms. This is the formation of chemical bonding, which is similar to the bright lines shown in Fig. 7.1. This means that the electronic wave function in a molecule can be LCAO. (3) As the simplest approximation, hydrogen-like AOs can be the candidates.

Therefore, the electronic wave function in a molecule is given as follows:

$$\psi = \sum c_i \phi_i$$

where ψ is the electronic wave function, ϕ_i is the hydrogen-like AO and c_i is the combinational coefficient. It can be assessed that electrons will stay longer on the atoms with larger combinational coefficients. For equivalent atoms, their coefficients are the same. Different coefficients of different atoms show the distinguishability with regard to electronic motion. Hydrogen-like AOs are the ones that we often encounter: *ns*, *np* (p_x, p_y, p_z) and *nd* (d_{xy}, d_{xz}, d_{yz}, $d_{x^2-y^2}$, d_{z^2})

7.3 Determination of hybridization coefficients

Atomic orbitals can form a molecular orbital with specific configuration through LCAO; this is called hybridization. For instance, for the configuration shown in Fig. 7.2, its molecular orbital can be given as follows:

$$\psi_1 = a_1 s + b_1 p_x$$
$$\psi_2 = a_2 s + b_2 p_x + c_2 p_y$$
$$\psi_3 = a_3 s + b_3 p_x + c_3 p_y$$

By orthogonality and normalization of the orbitals and considering the following properties,

$$\langle s|s\rangle = 1,\ \ \langle s|p_i\rangle = 0,\ \ \langle p_i|p_j\rangle = \delta_{ij}$$

together with the following symmetry consideration:

$$\sigma_{xz}\psi_2 = \psi_3$$

$$\begin{bmatrix} 1 & 0 & 0 \\ 0 & 1 & 0 \\ 0 & 0 & -1 \end{bmatrix} \begin{bmatrix} a_2 \\ b_2 \\ c_2 \end{bmatrix} = \begin{bmatrix} a_3 \\ b_3 \\ c_3 \end{bmatrix}$$

$$C_3\psi_2 = \psi_1$$

$$\begin{bmatrix} 1 & 0 & 0 \\ 0 & -\frac{1}{2} & \frac{\sqrt{3}}{2} \\ 0 & -\frac{\sqrt{3}}{2} & -\frac{1}{2} \end{bmatrix} \begin{bmatrix} a_2 \\ b_2 \\ c_2 \end{bmatrix} = \begin{bmatrix} a_1 \\ b_1 \\ 0 \end{bmatrix}$$

the coefficients can be determined as follows:

$$\begin{aligned} \psi_1 &= \tfrac{1}{\sqrt{3}}s + \sqrt{2/3}p_x \\ \psi_2 &= \tfrac{1}{\sqrt{3}}s - \tfrac{1}{\sqrt{6}}p_x + \tfrac{1}{\sqrt{2}}p_y \\ \psi_3 &= \tfrac{1}{\sqrt{3}}s - \tfrac{1}{\sqrt{6}}p_x + \tfrac{1}{\sqrt{2}}p_y \end{aligned}$$

Another alternative is by the group-theoretical method. Using $\{\psi_1, \psi_2, \psi_3\}$ as the basis, the characters under D_{3d} is given as follows:

D_{3d}	E	$2C_3$	$3C'_2$	σ_h	$2S_3$	$3\sigma_v$
χ_Γ	3	0	1	3	0	1

By reducing the representation, we have

$$\Gamma = A'_1 + E'$$

Note that x^2+y^2 and z^2 belong to A_1', and (x, y) and (x^2-y^2, xy) belong to E'. Hence, the hybridization is sp^2, sd^2, $d_{z^2}p^2$ or $d_{z^2}d^2$. The group-theoretical method can offer the symmetries, but not the coefficients. In most cases, this is just enough.

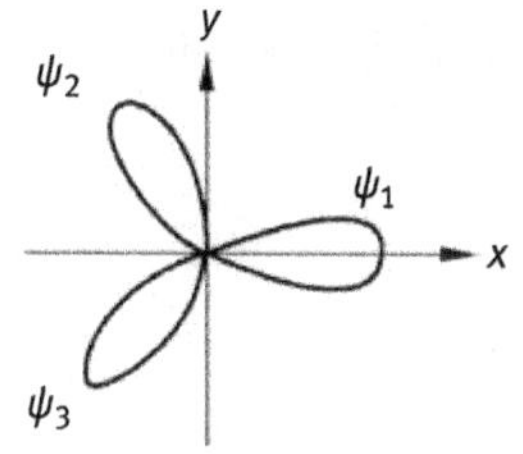

Fig. 7.2: The hybridization resulting in the D_{3d} configuration.

For the tetrahedral configuration shown in Fig. 7.3(a), we have

T_d	E	$8C_3$	$3C_2$	$6S_4$	$6\sigma_d$
χ_Γ	4	1	0	0	2

$\Gamma = A_1 + T_2$. Since $x^2+y^2+z^2$ belongs to A_1, and (x, y, z) and (xy, yz, xz) belong to T_2, the hybridization is sp^3 or sd^3.

For the square configuration of D_{4h} shown in Fig. 7.3(b), we have $\Gamma = B_{1g} + E_u + A_{1g}$. x^2-y^2 and (x, y) belong to B_{1g} and E_u, respectively, x^2+y^2 and z^2 belong to A_{1g} and the hybridization is dsp^2 or d^2p^2.

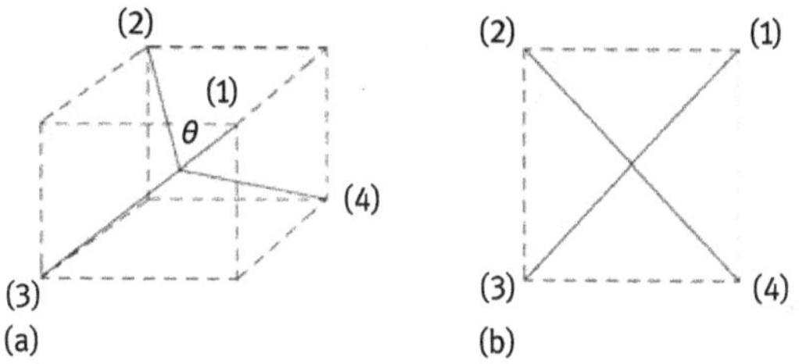

Fig. 7.3: The tetrahedral and square configurations.

7.4 Secular equation

Suppose that the system Hamiltonian is H. For any function ψ, it can be shown that

$$\int \psi^* H\psi / \int \psi^* \psi = \varepsilon \geq \varepsilon_0$$

where ε_0 is the energy of the ground state. Assuming $\Psi = \sum C_i \phi_i$, by variation, we require that $\partial\varepsilon/\partial C_k = 0$. Here, ε can be expressed as follows:

$$\varepsilon = \frac{\sum\limits_{i,j} C_i C_j H_{ij}}{\sum\limits_{i,j} C_i C_j S_{ij}}$$

with $H_{ij} = \int \phi_i^* H\phi_j$. When $i = j$, it is called the Coulomb energy, and when $i \neq j$, it is called the resonance energy. $S_{ij} = \int \phi_i^* \phi_j$, as $i = j$, $S_{ij} = 1$ and as $i \neq j$, it is called the overlap integral. Then we have

$$\partial\varepsilon/\partial C_k = (A - B)/C$$

with $A = \sum C_i C_j S_{ij} \partial/\partial C_k \sum C_i C_j H_{ij}$, $B = \sum C_i C_j H_{ij} \partial/\partial C_k \sum C_i C_j S_{ij}$ and $C = \left(\sum C_i C_j S_{ij}\right)^2$.

Note that

$$\partial/\partial C_k\Big(\sum_i C_i \sum_j C_j H_{ij}\Big) = \sum_{j\neq k} C_j H_{kj} + 2C_k H_{kk} + \sum_{i\neq k} C_i H_{ik} = 2\sum_i C_i H_{ki}$$

$$\partial/\partial C_k\Big(\sum_i C_i \sum_j C_j S_{ij}\Big) = 2\sum_i C_i S_{ki}$$

Hence,

$$\begin{aligned} 0 &= \sum C_i C_j S_{ij}\Big(2\sum C_i H_{ki}\Big) - \sum C_i C_j H_{ij}\Big(2\sum C_i S_{ki}\Big) \\ &= \sum C_i H_{ki} - \Big\{\sum C_i C_j H_{ij} / \sum C_i C_j S_{ij}\Big\} \sum C_i S_{ki} \end{aligned}$$

and $\sum_i C_i(H_{ki} - \varepsilon S_{ki}) = 0$

For this equation set to have a nontrivial solution, we need to consider the secular equation:

$$|H_{ki} - \varepsilon S_{ki}| = 0$$

or

$$\begin{vmatrix} H_{11} - \varepsilon S_{11} & H_{12} - \varepsilon S_{12}, & \cdots, & H_{1n} - \varepsilon S_{1n} \\ H_{21} - \varepsilon S_{21} & \cdots & & \cdots \\ \vdots & \vdots & & \vdots \\ H_{n1} - \varepsilon S_{n1} & \cdots & \cdots & H_{nn} - \varepsilon S_{nn} \end{vmatrix} = 0$$

7.5 Huckel approximation

Huckel molecular orbital (HMO) approximation tries to simplify the aforementioned equation. It includes the following:

1. When $H_{ii} = \alpha < 0$, the energy is that of $2p_\pi$ atomic orbital.
2. $H_{ij} = \beta < 0$ if i, j atoms are bonded.
3. $H_{ij} = 0$ if i, j atoms are not bonded.
4. $S_{ij} = \delta_{ij}$. This is an extreme approximation. If it is lessened, the approximation is called the extended HMO or EHMO.

In summary, HMO approximation is given as follows:

1. Write down the secular equation according to the molecular structure:

$$\begin{vmatrix} \alpha-\varepsilon & \beta_{12} & \cdots & \beta_{1n} \\ \beta_{21} & \alpha-\varepsilon & \cdots & \cdots \\ \vdots & \vdots & \ddots & \vdots \\ \beta_{n1} & \cdots & \cdots & \alpha-\varepsilon \end{vmatrix} = 0$$

2. Solve the secular equation.
3. Obtain ε and C_i.

Example 1: Ethylene $CH_2=CH_2$

For this molecule, we consider only the two $2p_z$ orbitals on its two atoms, as shown in Fig. 7.4.

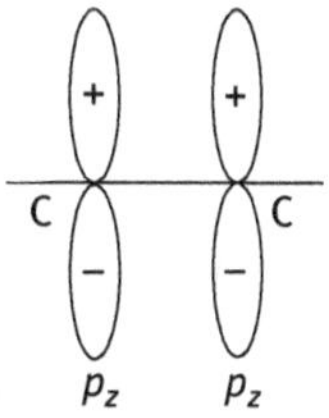

Fig. 7.4: The two $2p_z$ orbitals on the two atoms.

The secular equation is given as follows:

$$\begin{vmatrix} \alpha-\varepsilon & \beta \\ \beta & \alpha-\varepsilon \end{vmatrix} = 0$$

The solutions are $\varepsilon_1=\alpha+\beta$ and $\varepsilon_2=\alpha-\beta$. The wave functions are as follows:

$$\psi_1=\frac{1}{\sqrt{2}}(\phi_1+\phi_2),\quad \psi_2=\frac{1}{\sqrt{2}}(\phi_1-\phi_2)$$

Figure 7.5 shows the spatial configurations of the two MOs.

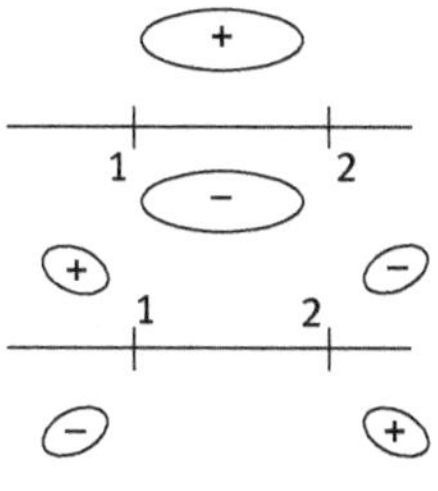

Fig. 7.5: The spatial configurations of the two MOs of ethylene.

We observe that at the low-energy level, its wave function possesses more density in between the two atoms for the bond formation. At the high-energy level, its electron density is less in between the two atoms. This weakens the bond formation and is called antibonding. Their energy levels are shown in Fig. 7.6, where the arrows indicate the electrons. For each level, there can be at most two occupying paired electrons. For the ethylene model, there are two electrons, and the total energy is $2(\alpha+\beta)$. When compared to the energy before bonding, 2α, the stability energy is $2\beta < 0$. Thus, the system is stable.

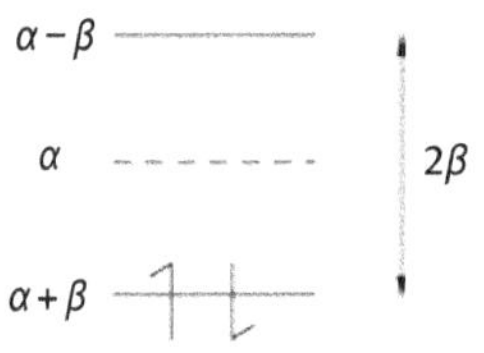

Fig. 7.6: The energy levels for the ethylene model.

Example: 2 $[C-C-C]^+$ **(allyl)**
The structure of this system is given as shown in Fig.7.7:

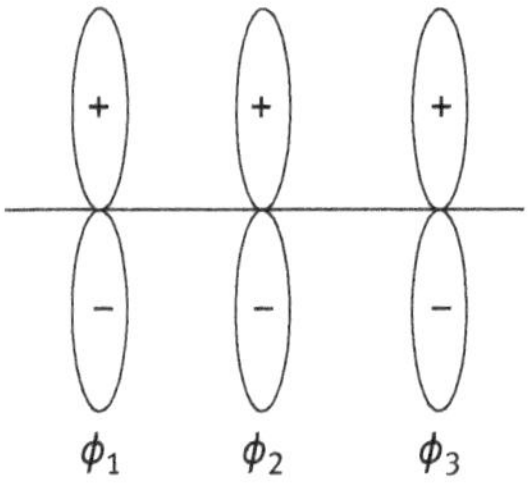

Fig. 7.7: The allyl system.

The secular equation is given as follows:

$$\begin{vmatrix} \alpha-\varepsilon & \beta & 0 \\ \beta & \alpha-\varepsilon & \beta \\ 0 & \beta & \alpha-\varepsilon \end{vmatrix} = 0$$

For convenience, we define $x = (\alpha-\varepsilon)/\beta$, then

$$\begin{vmatrix} x & 1 & 0 \\ 1 & x & 1 \\ 0 & 1 & x \end{vmatrix} = 0$$

The solutions are $\varepsilon_1 = \alpha + \sqrt{2}\beta$, $\varepsilon_2 = \alpha$ and $\varepsilon_3 = \alpha - \sqrt{2}\beta$. The lowest and highest levels are the bonding and antibonding orbitals, respectively. The middle level does not contribute to the bond formation and is called nonbonding. Their wave functions are given as follows:

$$\psi_1 = \frac{1}{2}(\phi_1 + \sqrt{2}\phi_2 + \phi_3),$$

$$\psi_2 = \frac{1}{\sqrt{2}}(\phi_1 - \phi_3),$$

$$\psi_3 = \frac{1}{2}(\phi_1 - \sqrt{2}\phi_2 + \phi_3)$$

The node number in the wave function increases with increase in the level. For the lowest bonding orbital, there are no nodes. For the nonbonding orbital, the central atom does not contribute to the system. The antibonding orbital possesses two nodes. The energy-level structures of the neutral, cation and anion species are shown in Fig. 7.8. These systems are stable. The electrons in the neutral and anion configurations do not affect the stability since they are in the nonbonding orbital.

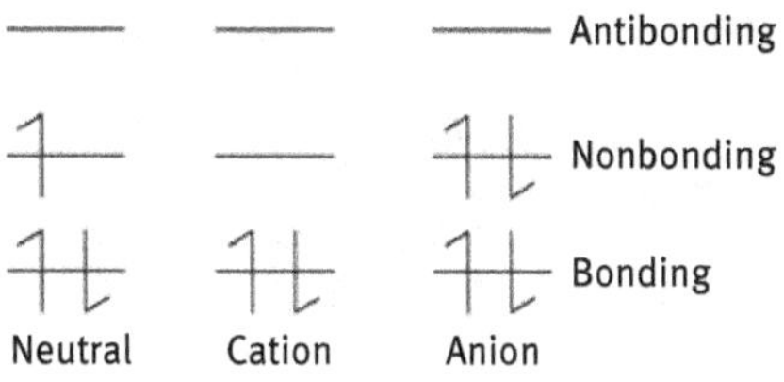

Fig. 7.8: The energy-level structures of the neutral, cation and anion allyl systems.

Example 3: Butadiene $[C=C-C=C]$
The structure is shown in Fig. 7.9.

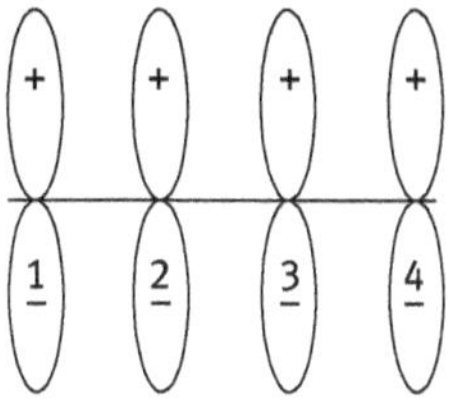

Fig. 7.9: The structure of butadiene.

The secular equation is given as follows:

$$\begin{vmatrix} x & 1 & 0 & 0 \\ 1 & x & 1 & 0 \\ 0 & 1 & x & 1 \\ 0 & 0 & 1 & x \end{vmatrix} = 0$$

Energies and wave functions are given as follows:

ε	ϕ_1	ϕ_2	ϕ_3	ϕ_4	
$\alpha + 1.618\beta$	0.371	0.6	0.6	0.371	ψ_1
$\alpha + 0.618\beta$	0.6	0.371	−0.371	−0.6	ψ_2
$\alpha - 0.618\beta$	0.6	−0.371	−0.371	0.6	ψ_3
$\alpha - 1.618\beta$	0.371	−0.6	0.6	−0.371	ψ_4

The node structures are shown in Fig. 7.10. The levels of higher energy possess more nodes.

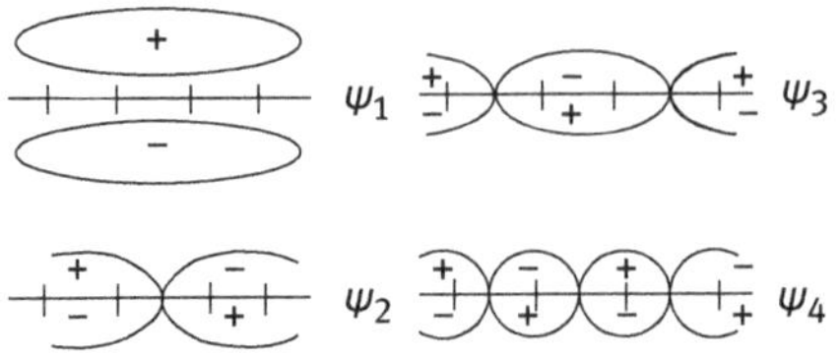

Fig. 7.10: The node structures of the butadiene system.

The π electron density is defined as $q_r = \sum_j n_j C_{jr}{}^2$. The bond order is $b_{rs} = \sum_j n_j C_{jr} C_{js}$. Here j indicates the electron-occupied orbital and r is for the atom.

For this system, we have $q_1 = q_2 = q_3 = q_4 = 1$, $b_{12} = b_{34} = 0.894$ and $b_{23} = 0.447$. This is consistent with the usual formula of $[C = C - C = C]$.

7.6 Symmetry and group method

If there is symmetry, the group method can be employed to simplify the treatment and to classify/label the orbitals and states. For instance, in the case of ethylene, we can choose $\psi_1 = \frac{1}{\sqrt{2}}(\phi_1 + \phi_2)$ and $\psi_2 = \frac{1}{\sqrt{2}}(\phi_1 - \phi_2)$ to construct the secular equation as follows:

$$\begin{array}{c|cc|l} & \psi_1 & \psi_2 & \\ \psi_1 & \alpha + \beta - \varepsilon & 0 & \\ \psi_2 & 0 & \alpha - \beta - \varepsilon & \end{array} = 0$$

This equation is already diagonal. How to choose such $\{\psi_1, \psi_2\}$ as the basis? This can be done by the group theory. The chosen wave functions have to be the basis of the irreducible representations. For butadiene, if the configuration is as shown in Fig. 7.11 of the C_{2v} group, it can be checked whether the wave functions are of the symmetries as $\psi_1(B_1), \psi_2(A_2), \psi_3(B_1), \psi_4(A_2)$.

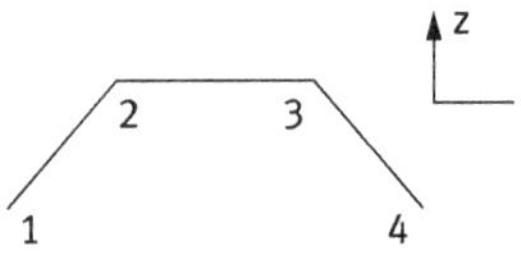

Fig. 7.11: The configuration of butadiene of C_{2v} symmetry.

These symmetrized wave functions can be employed to construct the secular equation. Since H is of total symmetry, if the product of Γ_i and Γ_j, which are the symmetries of ψ_i, ψ_j, does not contain the total symmetry, then $\langle \psi_i H \psi_j \rangle$ will be zero. For the overlap integral $\langle \psi_i \psi_j \rangle$, this criterion also applies.

For example, for butadiene, employ $(\phi_1, \phi_2, \phi_3, \phi_4)$ as the basis. The characters are given as follows:

$$\chi_E = 4,\ \chi_{C_2} = 0,\ \chi_{\sigma_V} = 0,\ \chi_{\sigma_V'} = -4$$

By reduction, $\Gamma = 2A_2 + 2B_1$. The wave functions of A_2, B_1 symmetries are given as follows:

$$\Phi_1(A_2) = 1/\sqrt{2}(\phi_1 - \phi_4),\quad \Phi_2(A_2) = 1/\sqrt{2}(\phi_2 - \phi_3)$$

$$\Phi_3(B_1) = 1/\sqrt{2}(\phi_1 + \phi_4),\quad \Phi_4(B_1) = 1/\sqrt{2}(\phi_2 + \phi_3)$$

By these wave functions, the secular equation is written as follows:

$$\begin{array}{c} \\ A_2\left\{\begin{array}{c}\\ \\ \end{array}\right. \\ B_1\left\{\begin{array}{c}\\ \\ \end{array}\right. \end{array}\begin{array}{c}\overbrace{\hphantom{\alpha-\varepsilon \quad \alpha-\beta-\varepsilon}}^{A_2}\overbrace{\hphantom{\alpha-\varepsilon \quad \alpha+\beta-\varepsilon}}^{B_1} \\ \begin{vmatrix} \alpha-\varepsilon & \beta & 0 & 0 \\ \beta & \alpha-\beta-\varepsilon & 0 & 0 \\ 0 & 0 & \alpha-\varepsilon & \beta \\ 0 & 0 & \beta & \alpha+\beta-\varepsilon \end{vmatrix}\end{array} = 0$$

Note that the elements between different symmetries are zero. In principle, we can employ any kind of wave functions to construct the secular equation. However, only the equations with proper symmetry will make the calculation simple.

7.7 Correlation

As the configuration of the system transforms, there is a change in its symmetry as well. Often, the transformation is a minor perturbation such that there is a correlation among the symmetries of the wave functions before and after the transformation. For instance, for the transformation from the planar square to the nonplanar configurations as shown in Fig. 7.12, the symmetry changes from D_{4h} to D_{2d}. Their irreducible representations are $A_{2u} + B_{1u} + E_g$ and $A_2 + B_1 + E$, respectively. The wave functions of D_{4h} are $\psi_{A_{2u}} = 1/2(\phi_1 + \phi_2 + \phi_3 + \phi_4)$, $\psi_{E_g} = 1/\sqrt{2}(\phi_1 - \phi_3)$, $\psi'_{E_g} = 1/\sqrt{2}(\phi_2 - \phi_4)$, $\psi_{B_{1u}} = 1/2(\phi_1 - \phi_2 + \phi_3 - \phi_4)$

The wave functions of D_{2d} are the same as above, but with different symmetries. Their correlations are shown in Fig. 7.13.

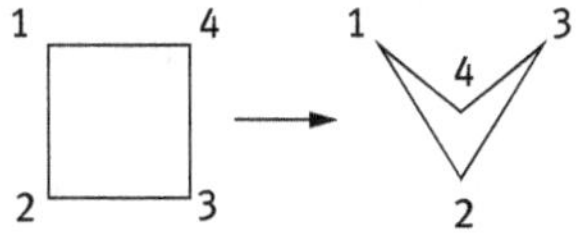

Fig. 7.12: The transformation from the planar square to the non-planar configurations.

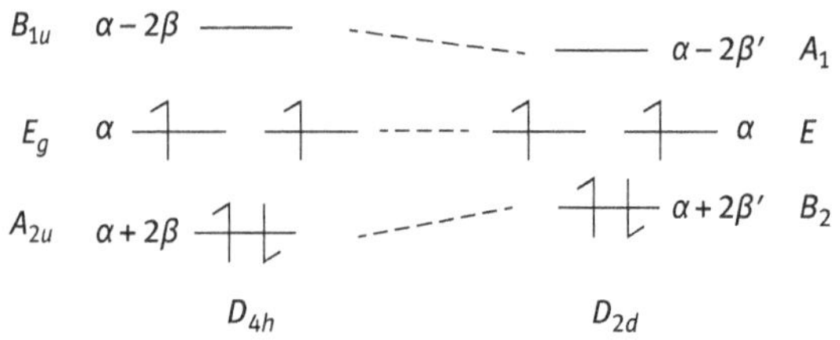

Fig. 7.13: The correlations of the levels of different symmetries between D_{4h} and D_{2d}.

By the concept of correlation, Hoffmann and Woodward proposed the principles to predict chemical reactions.

7.8 Modifications of HMO

HMO accesses the physical core and is very successful. On the basis of this, some modifications were made:

(1) For α, different atoms and sites are assigned with different values.
(2) For β, for the nonbonded case, nonzero value is assigned but is smaller than that for the bonded case.
(3) For $S_{ij} = \delta_{ij}$, without this extreme approximation, instead, its value based on the atomic wave functions is calculated. This is known as the extended HMO (EHMO).
(4) Besides the π orbital, σ orbital along the bond is considered. The corresponding β is assigned with a different value.
(5) HMO is a one-electron approximation. The molecular wave function is the product of the electronic wave functions in different orbitals: $\psi_{HMO} = \prod_i \psi_i(n_i)$. Such a wave function does not satisfy the Pauli principle that it has to change its sign as the coordinates of any two electrons are exchanged. HMO-based wave function is, therefore, modified to satisfy the principle.
(6) HMO is a one-electron approximation. The interelectronic repulsive interaction is not considered. To overcome this issue, *configuration interaction* is introduced.

7.9 Electronic transition and selection rules

The electronic transition between ψ_n and ψ_m requires that $\langle\psi_n\mu\psi_m\rangle \neq 0$ where the symmetry of μ is $\Gamma_x, \Gamma_y, \Gamma_z$. Hence, for $\langle\psi_n\mu\psi_m\rangle \neq 0$, it is required that the reduction of $\Gamma(\psi_n) \bullet \Gamma(\psi_m)$ has to contain $\Gamma_x, \Gamma_y, \Gamma_z$. Since the ground state is of total symmetry, the selection rule is as follows: the excited state has to be of $\Gamma_x, \Gamma_y, \Gamma_z$ symmetries. As there is inversion symmetry, $\Gamma_x, \Gamma_y, \Gamma_z$ is of u symmetry. Then, only those excited states of this symmetry will be optically active. This is known as the Laporte rule.

For example, Fig. 7.14 shows the C_{2h} system constituted by $2p_z$ orbitals.

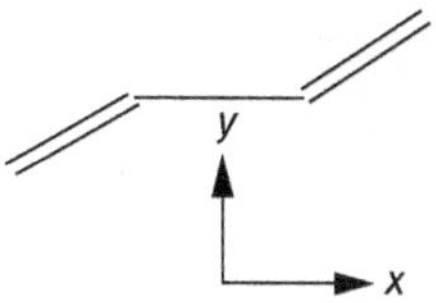

Fig. 7.14: The C_{2h} system constituted by $2p_z$ orbitals.

The electronic transition is shown in Fig. 7.15.

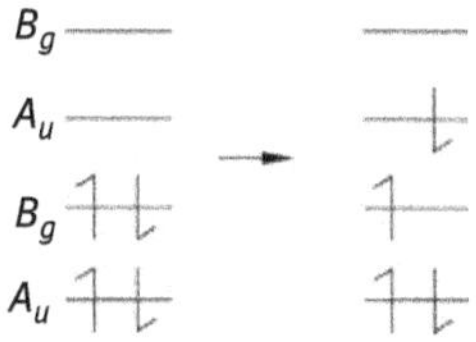

Fig. 7.15: The electronic transition of the C_{2h} system.

Here, the spin conservation is considered. The symmetry of the initial state is $A_u \bullet A_u \bullet B_g \bullet B_g = A_g$ and that of the final state is $A_u \bullet A_u \bullet B_g \bullet A_u = B_u$. Hence, the photon has to be of symmetry $\Gamma_x = B_u$, that is, only light with x polarization is active. From Fig. 7.14, it can be seen that only the electric field of x polarized light can drift the electrons along the system and induce the transition.

7.10 Comments

HMO is a very simple and crude approximation. However, the symmetries of the electronic orbitals based on HMO are very accurate. Symmetry regulates the electronic motion. Even though today's software is very advanced, HMO still holds its advantages. Its importance cannot be underestimated.

Exercises

7.1 For this system, write the wave functions, energy levels and their symmetries.

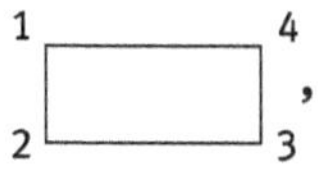

$, \beta_{12} > \beta_{14}, S_{12} > S_{14}$

7.2 For the system in $(x–y)$ plane 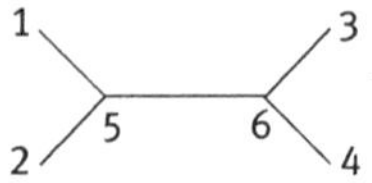

with P_z on each site: ϕ_1, ϕ_2, ϕ_3, ϕ_4, ϕ_5, ϕ_6.

The in-plane coordinates are 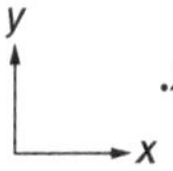

.z axis is vertical to the plane.

(a) Show the symmetries of its MO under D_{2h}.
(b) Write the MO LCAO of ϕ_i.
(c) For the neutral case, write the symmetries of its ground and two low excited states.
(d) Is the transition from the ground state to the above-mentioned two excited states optically active? By what polarized light?

References

[1] Levine IN. Quantum Chemistry. Boston: Allyn and Bacon Inc., 1974.
[2] Ballhausen CJ, Gray HB. Molecular Orbital Theory. 1964.
[3] Murrell JN, Harget AJ.Semi-empirical Self-consistent-field Molecular Orbital Theory. London-New York-Sydney-Toronto: Wiley-Interscience, 1972.

8 Raman effect

8.1 Scattering

In the previous chapters, we have discussed about the light interaction with molecules where the interaction is mediated by the coupling of the electric field of light with the dipoles (permanent or vibrationally induced) of molecules. This results in the light absorption in IR spectral region. In this chapter, we will introduce another important effect by the light interaction with molecules – the Raman effect. The concept of the Raman effect is that as a photon (scattering light) with high frequency (compared with the vibration) is absorbed by the molecule, the electrons in the molecule are excited/disturbed (not necessarily to the eigenstate, often called the virtual state). The excited/disturbed electrons are very unstable and immediately relax to the ground state, with the emission of a secondary photon (scattered light) whose energy is lesser or larger than the first absorbed photon, depending on whether the energy of the excited/disturbed electrons is transferred to the molecular internal motion, such as vibration, or the energy of the molecular internal motion is transferred to the electronic motion and combined together with the emitted photon. The energy transfer is via the *vibronic* coupling between the electronic and vibrational motions. Raman effect offers more molecular information than that obtained by the IR absorption process. However, in the scattering, there is the case that involves no exchange of energy between the light and molecule; this is called the Rayleigh process.

The emitted scattered light can be considered as the radiation by an oscillating dipole with the moment $\mu_Z = \mu_Z^0 \cos \omega t$. According to the classical dynamics, the radiation intensity along $\theta = 90°$ plane (vertical to the dipole) is given as follows:

$$I(\nu) = \frac{2\pi^3 \nu^4}{c^3} \left(\mu_Z^0\right)^2$$

where c is the speed of light. The scattering configuration is shown in Fig. 8.1.

Since the molecule is not always spherical, the induced dipole μ_{ind} is not always along the electric field of the scattering light. Hence, we have the following equation:

$$\mu_{\text{ind}} = \begin{bmatrix} \mu_X \\ \mu_Y \\ \mu_Z \end{bmatrix} = \begin{bmatrix} \alpha_{XX} & \alpha_{XY} & \alpha_{XZ} \\ \alpha_{YX} & \alpha_{YY} & \alpha_{YZ} \\ \alpha_{ZX} & \alpha_{ZY} & \alpha_{ZZ} \end{bmatrix} \begin{bmatrix} \varepsilon_X \\ \varepsilon_Y \\ \varepsilon_Z \end{bmatrix}$$

where α_{ij} is the proportionality of the electric moment along the i direction induced by the electric field along the j direction of the scattering light. α_{ij} is called the polarizability.

https://doi.org/10.1515/9783110625097-008

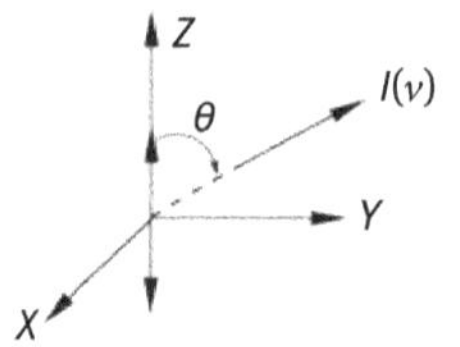

Fig. 8.1: The scattering configuration.

Now, the electrons in the molecule are under two forces: (1) one is the attraction by the nuclei and the other is (2) the exertion by the electric field of the scattering light. (Of course, there is the repulsion among the electrons. Since we consider the electrons as a whole, we will not consider this repulsion.) Polarizability is an expression of the balance of these two forces. If the attraction by the nuclei is large, the polarizability is small. Conversely, if the electrons are more influenced by the electric field of the light, then the polarizability is larger.

We note that the dimension of polarizability is volume. (In the case of two or one dimension, its dimension is the area or length.) Its physical content is the space of electrons that are bound by the molecular nuclei. Polarizability is roughly proportional to the electron number in the molecule.

The scattering configuration is often described by the notation: $A(B\ C)D$, where A is the direction of the scattering light, B the direction of the electric field of the scattering light, C the direction of the electric field of the scattered light and D the direction of the scattered light. For instance, for the configuration shown in Fig. 8.2, the notation of the scattering configuration is $Y(ZY)X$.

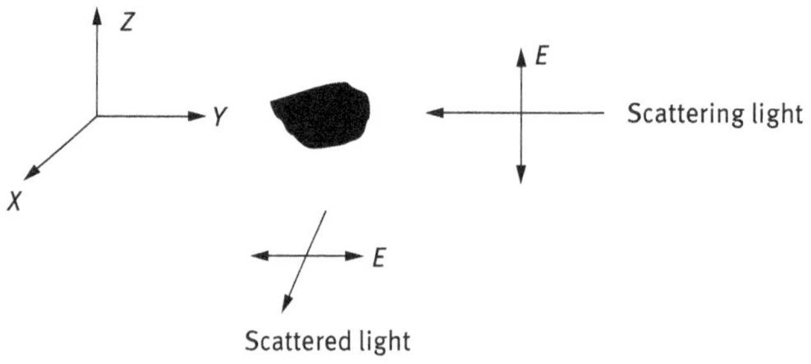

Fig. 8.2: The scattering configuration.

In such a condition, the scattered intensity is

$$I_{Y(ZY)X} = \frac{2\pi^3 \nu^4}{c^3} \left(\mu_Y^0\right)^2$$

where the dipole μ_Y^0 in the Y direction originates from ε_Z in the Z direction:

$$\mu_Y^0 = \alpha_{YZ}\varepsilon_Z$$

Then

$$I_{Y(ZY)X} = \frac{2\pi^3\nu^4}{c^3}\alpha_{YZ}^2\varepsilon_Z^2$$

Since the scattering light intensity I_0 is

$$I_0 = \frac{c}{8\pi}\varepsilon_Z^2$$

we have $I_{Y(ZY)X} = \frac{16\pi^4\nu^4}{c^4}\alpha_{YZ}^2 I_0$

For the $Y(X_Z^Y)X$ configuration, we have

$$\begin{aligned} I_{Y(X_Z^Y)X} &= \tfrac{2\pi^3\nu^4}{c^4}\left[\left(\mu_Y^0\right)^2 + \left(\mu_Z^0\right)^2\right] \\ &= \tfrac{16\pi^4\nu^4}{c^4}\left[\alpha_{YX}^2 + \alpha_{ZX}^2\right] I_0 \end{aligned}$$

where $\{XYZ\}$ is the laboratory coordinates. As $i \neq j$, α_{ij} is nonzero in general. We may choose the principal coordinates $\{xyz\}$ on the molecule such that

$$\alpha_{ij} = \alpha_i\delta_{ij} \quad \delta_{ij} = \begin{cases} 0 & i \neq j \\ 1 & i = j \end{cases}$$

$\alpha_{FF'}$ in $\{XYZ\}$ can be related to α_i in $\{xyz\}$ as follows:

$$\alpha_{FF'} = \sum \Phi_{Fi}\Phi_{F'i}\alpha_i$$

where Φ_{Fi} is the directional cosine between the two coordinates.

For molecules in free space without fixed orientation, averaging is required, which can be given as follows:

$$\overline{\alpha_{FF'}^2} = \sum_i \overline{\Phi_{Fi}^2\ \Phi_{F'i}^2}\alpha_i^2 + 2\sum_{i<j}\overline{\Phi_{Fi}\ \Phi_{F'i}\Phi_{Fj}\Phi_{F'j}}\alpha_i\ \alpha_j \tag{8.1}$$

which can be calculated as follows [1]:

$$\begin{cases} \frac{1}{5}\sum\limits_i \alpha_i^2 + \frac{2}{15}\sum\limits_{i<j}\alpha_i\alpha_j & F = F' \\ \frac{1}{15}\sum\limits_i \alpha_i^2 - \frac{1}{15}\sum\limits_{i<j}\alpha_i\alpha_j & F \neq F' \end{cases} \tag{8.2}$$

Then for $I_{Y(X_Z^Y)X}$ and $I_{Y(ZY)X}$, we have

$$I_{Y(X^Y_Z)X} = \frac{2}{15}\frac{16\pi^4 v^4 I_0}{c^4}\left(\sum_i \alpha_i^2 - \sum_{i<j}\alpha_i\alpha_j\right)$$
$$I_{Y(ZY)X} = \frac{1}{15}\frac{16\pi^4 v^4 I_0}{c^4}\left(\sum_i \alpha_i^2 - \sum_{i<j}\alpha_i\alpha_j\right)$$

Introducing $\bar{\alpha}$ and β^2,

$$\bar{\alpha} = \frac{1}{3}(\alpha_1 + \alpha_2 + \alpha_3)$$
$$\beta^2 = \frac{1}{2}\left[(\alpha_1 - \alpha_2)^2 + (\alpha_2 - \alpha_3)^2 + (\alpha_3 - \alpha_1)^2\right]$$

where $\bar{\alpha}$ is the averaged polarizability (along the three principal axes) and β is the deviation of the polarizabilities from a spherical distribution. In terms of $\bar{\alpha}$ and β, we have the following:

$$I_{Y(X^Y_Z)X} = \frac{16\pi^4 v^4 I_0}{c^4}\frac{2\beta^2}{15}$$
$$I_{Y(ZY)X} = \frac{16\pi^4 v^4 I_0}{c^4}\frac{3\beta^2}{45}$$

The intensity ratio of the scattered light along the X direction with polarization along the Y direction to that along the Z direction when the scattering light is along the Y direction is called the depolarization ratio and is denoted by ρ_l. It shows the *amount of rotation* of the polarization from Z to Y directions after scattering. If the scattering light is unpolarized along the X and Z directions, then the depolarization ratio is denoted by ρ_n. It shows the inhomogeneous distribution of the polarization of the scattered light when originally the scattering light is homogeneously polarized. The background of this phenomenon is the nonspherical distribution of polarizability, or the molecule is not spherical. After scattering by a molecule, the scattered light then contains the structural information of the molecule. This is a very important concept.

After simple calculation, we have the following:

$$\rho_l = \frac{I_{Y(ZY)X}}{I_{Y(ZZ)X}} = \frac{\alpha_{YZ}^2}{\alpha_{ZZ}^2} = \frac{3\beta^2}{45\bar{\alpha}^2 + 4\beta^2}$$
$$\rho_n = \frac{I_{Y(^X_Z Y)X}}{I_{Y(^X_Z Z)X}} = \frac{6\beta^2}{45\bar{\alpha}^2 + 7\beta^2}$$

Their ranges are as follows:

$$0 \leqq \rho_l \leqq \frac{3}{4}$$

$$0 \leqq \rho_n \leqq \frac{6}{7}$$

corresponding to β=0, or $\bar{\alpha}^2 \gg \beta^2$ and $\bar{\alpha}^2 \ll \beta^2$, respectively. When a molecule is spherical, β=0, the depolarization ratio is 0. On the contrary, if a molecular shape deviates much from sphere (such as linear), then the ratios reach their maxima. Hence, by measuring the depolarization ratios, a lot can be known about the molecular shape.

8.2 Raman effect

Suppose that the electric field of the scattering light is

$$\varepsilon = \varepsilon_0 e^{i\omega t}$$

and the normal coordinate Q_k is with angular frequency ω_k:

$$Q_k = Q_k^0 \cos \omega_k t$$
$$= Q_k^0 \frac{1}{2} \left[e^{i\omega_k t} + e^{-i\omega_k t}\right]$$

Furthermore, the electronic polarizability α is dependent on Q_k. By expanding α in terms of Q_k (for convenience, in the following, the subscript of polarizability will be omitted without confusion), we have the following:

$$\alpha = \alpha_0 + \sum_k (\partial\alpha/\partial Q_k)_0 Q_k + \ldots$$

Under the influence of the scattering light, the induced electric dipole μ is given as follows:

$$\mu = \alpha\varepsilon$$
$$= \{\alpha_0 + \sum_k (\partial\alpha/\partial Q_k)_0 Q_k^0 \frac{1}{2}[e^{i\omega_k t} + e^{-i\omega_k t}]\} \cdot \varepsilon_0 e^{i\omega t}$$
$$= \alpha_0 \varepsilon_0 e^{i\omega t} + \frac{1}{2}\sum_k (\partial\alpha/\partial Q_k)_0 Q_k^0 \varepsilon_0 \left[e^{i(\omega+\omega_k)t} + e^{i(\omega-\omega_k)t}\right]$$

where μ contains $e^{i\omega t}$, $e^{i(\omega+\omega_k)t}$ and $e^{i(\omega-\omega_k)t}$. This shows that the scattered light possesses the frequencies $\omega \pm \omega_k$ besides ω. Figure 8.3 shows its spectrum.

The line with a frequency ω is the Rayleigh line. Its intensity is related to α_0, which is the molecular property as a whole and is independent of vibration. The

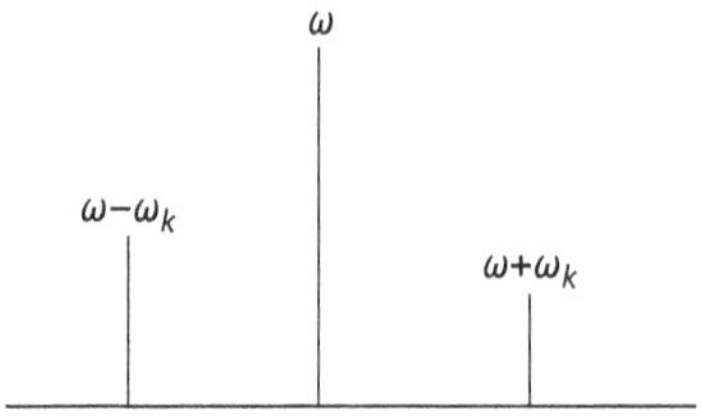

Fig. 8.3: The Rayleigh and Raman lines. Those with $\omega - \omega_k$ and $\omega + \omega_k$ are the Stokes and anti-Stokes lines, respectively.

Rayleigh process involves no nuclear motion. The lines with frequencies $\omega \pm \omega_k$ are the Raman lines. The one with $\omega - \omega_k$ are the Stokes lines and the ones with $\omega + \omega_k$ are the anti-Stokes lines. The former can be considered as that the energy of one vibrational quantum is absorbed from the scattering light by the molecule while the latter is that one vibrational quantum is transferred to the scattered light from the molecule. Their intensity ratio is (no other factors are considered for the time being) given as follows:

$$\frac{I_{\text{Stokes}}}{I_{\text{antiStokes}}} = \frac{(\omega - \omega_k)^4 N_{V_k=0}}{(\omega + \omega_k)^4 N_{V_k=1}}$$

Here, N_{V_k} is the molecular population with vibrational quantum V_k. Although $(\omega - \omega_k)^4 < (\omega + \omega_k)^4$, the population of the ground state is much larger:

$$N_{V_k=0} \gg N_{V_k=1}$$

Hence, the Stokes line is stronger than the anti-Stokes line.

In conclusion, we emphasize that α_0 is the static polarizability when the molecular motion is *frozen*. It shows roughly the molecular dimension and shape. $(\partial\alpha/\partial Q_k)_0$ shows the interaction between the electronic and nuclear motions.

8.3 Quantal treatment

In Chapter 1, we have shown that under the effect of light (Hamiltonian is H'), the system wave function Ψ_a can be expressed as follows:

$$\Psi_a = \sum_n C_{n,a} \Psi_n^{(0)} e^{-iE_n t/\hbar}$$

where $\Psi_n^{(0)}$ is the one when there is no light. Initially, the system state is $|a\rangle$, then

$$C_{a,a}(t) = 1 + \frac{1}{i\hbar}\int_0^t \langle a|H'(t')|a\rangle \, dt' \tag{8.3}$$

$$C_{n\neq a,a}(t) = \frac{1}{i\hbar}\int_0^t \langle n|H'(t')|a\rangle e^{-\frac{i}{\hbar}(E_a - E_n)t'}dt' \tag{8.4}$$

$H'(t)$ can be written as follows:

$$H'(t) = Ve^{i\omega t} + V^{*}e^{-i\omega t} \tag{8.5}$$

where $V = \mu\varepsilon$, $V^{*} = \mu\varepsilon^{*}$. From eqs. (8.4) and (8.5), we have

$$C_{n,a}(t) = \frac{-\langle n|\mu\varepsilon|a\rangle}{\hbar}\frac{1-e^{-i(\omega_{an}-\omega)t}}{\omega_{an}-\omega} - \frac{\langle n|\mu\varepsilon^{*}|a\rangle}{\hbar}\frac{1-e^{-i(\omega_{an}+\omega)t}}{\omega_{an}+\omega} \tag{8.6}$$

where

$$\left(E_i - E_j\right)/\hbar = \omega_{ij},\ E_i/\hbar = \omega_i \tag{8.7}$$

We neglect the time-independent terms, eq. (8.6) is given as follows:

$$C_{n,a}(t) \propto \frac{\frac{1}{\hbar}\langle n|\mu\varepsilon|a\rangle e^{-i(\omega_{an}-\omega)t}}{\omega_{an}-\omega} + \frac{\frac{1}{\hbar}\langle n|\mu\varepsilon^{*}|a\rangle e^{-i(\omega_{an}+\omega)t}}{\omega_{an}+\omega} \tag{8.8}$$

On the other hand, under the light effect, suppose that $C_{a,a}(t) \approx 1$, then Ψ_a can be written as follows:

$$\Psi_a = \Psi_a^{(0)}e^{-i\omega_a t} + \sum_{n\neq a} C_{n,a}\Psi_n^{(0)}e^{-i\omega_n t} \tag{8.9}$$

Now we can calculate the transition-induced dipole moment $\left(\mu_{\text{ind}}\right)_{ab}$:

$$\begin{aligned}\left(\mu_{\text{ind}}\right)_{ab} &= \langle a|\mu|b\rangle\\ &= \Big\langle \Psi_a^{(0)}e^{-i\omega_a t} + \sum_{n\neq a} C_{n,a}\Psi_n^{(0)}e^{-i\omega_n t}|\mu|\\ &\qquad \Psi_b^{(0)}e^{-i\omega_b t} + \sum_{m\neq b} C_{m,b}\Psi_m^{(0)}e^{-i\omega_m t}\Big\rangle\end{aligned} \tag{8.10}$$

For convenience, we define $|i_0\rangle \equiv |\Psi_i^{(0)}\rangle$. Equation (8.10) is shortened to

$$\begin{aligned}\left(\mu_{\text{ind}}\right)_{ab} &= \langle a_0|\mu|b_0\rangle e^{-i\omega_{ba}t} + \sum_{n\neq a} C_{n,a}^{*}\langle n_0|\mu|b_0\rangle e^{-i\omega_{bn}t}\\ &\quad + \sum_{m\neq b} C_{m,b}\langle a_0|\mu|m_0\rangle e^{-i\omega_{ma}t} + \ldots\end{aligned} \tag{8.11}$$

Neglecting the high-order terms and substituting $C_{n,i}(t)$ into eq. (8.11), we have the following:

$$\frac{1}{\hbar}\sum_{n\neq a}\left[\frac{\langle a_0|\mu\varepsilon^{*}|n_0\rangle e^{i(\omega_{an}-\omega)t}}{\omega_{an}-\omega} + \frac{\langle a_0|\mu\varepsilon|n_0\rangle e^{i(\omega_{an}+\omega)t}}{\omega_{an}+\omega}\right]\cdot\langle n_0|\mu|b_0\rangle e^{-i\omega_{bn}t}$$

$$
\begin{aligned}
&+\frac{1}{\hbar}\sum_{m\neq b}\left[\frac{\langle m_0|\mu\varepsilon|b_0\rangle e^{-i(\omega_{bm}-\omega)t}}{\omega_{bm}-\omega}+\frac{\langle m_0|\mu\varepsilon^*|b_0\rangle e^{-i(\omega_{bm}+\omega)t}}{\omega_{bm}+\omega}\right] \\
&\cdot\langle a_0|\mu|m_0\rangle e^{-i\omega_{ma}t}+\langle a_0|\mu|b_0\rangle e^{i\omega_{ab}t} \\
&=\frac{1}{\hbar}\left\{\sum_{n\neq a}\frac{\langle a_0|\mu\varepsilon|n_0\rangle\langle n_0|\mu|b_0\rangle}{\omega_{an}+\omega}+\sum_{m\neq b}\frac{\langle a_0|\mu|m_0\rangle\langle m_0|\mu\varepsilon|b_0\rangle}{\omega_{bm}-\omega}\right\}\cdot e^{i(\omega_{ab}+\omega)t} \\
&+\frac{1}{\hbar}\left\{\sum_{n\neq a}\frac{\langle a_0|\mu\varepsilon^*|n_0\rangle\langle n_0|\mu|b_0\rangle}{\omega_{an}-\omega}+\sum_{m\neq b}\frac{\langle a_0|\mu|m_0\rangle\langle m_0|\mu\varepsilon^*|b_0\rangle}{\omega_{bm}+\omega}\right\} \\
&\cdot e^{i(\omega_{ab}-\omega)t}+\langle a_0|\mu|b_0\rangle e^{i\omega_{ab}t}
\end{aligned}
\tag{8.12}
$$

The first term of eq. (8.12) corresponds to the scattered light with $\omega_{ab}+\omega$. As $E_a > E_b$, it is the anti-Stokes scattering, while for $E_a < E_b$, it is the Stokes scattering. The second term is also the Raman scattering but with opposite situation.

The ρ component of $(\mu_{\text{ind}})_{ab}$ is

$$
\sum_{\sigma=x,y,z}(\alpha_{\rho\sigma})_{ab}\varepsilon_\sigma \tag{8.13}
$$

and

$$
\mu\varepsilon=\sum_\sigma\mu_\sigma\varepsilon_\sigma \tag{8.14}
$$

From eq. (8.12), we have the quantal expression for the transition from state a to b as follows:

$$
(\alpha_{\rho\sigma})_{ab}=\frac{1}{\hbar}\left\{\sum_{n\neq a}\frac{\langle a_0|\mu_\sigma|n_0\rangle\langle n_0|\mu_\rho|b_0\rangle}{\omega_{an}+\omega}+\sum_{m\neq b}\frac{\langle a_0|\mu_\rho|m_0\rangle\langle m_0|\mu_\sigma|b_0\rangle}{\omega_{bm}-\omega}\right\}
$$

where μ_ρ and μ_σ are the polarizations of the photons involved in the two-photon process of the Raman scattering. The symmetry of μ is $\{x, y, z\}$. From the aforementioned expression, we know that the symmetry of $\alpha_{\rho\sigma}$ is the same as $\rho\sigma$.

8.4 Selection rules

From the initial state $|I\rangle$ to the final state $|F\rangle$, the Raman process depends on whether the transition integral

$$
\begin{aligned}
&\langle F|\mu_{\text{ind}}|I\rangle \\
&=\langle F|\alpha\varepsilon|I\rangle \\
&\propto\langle F|\alpha|I\rangle
\end{aligned}
$$

is zero or not.

The states are the product of the vibrational $|V\rangle$ and rotational $|R\rangle$ wave functions:

$$|I\rangle = |V\rangle|R\rangle = |VR\rangle$$
$$|F\rangle = |V'\rangle|R'\rangle = |V'R'\rangle$$

Since

$$\alpha_{FF'} = \sum_{gg'} \Phi_{Fg}\Phi_{F'g'}\alpha_{gg'}$$

and $\alpha_{gg'}$ is only dependent on vibrational coordinates, then

$$\begin{aligned}&\langle F|\alpha_{FF'}|I\rangle \\ &= \langle V'R'|\alpha_{FF'}|VR\rangle \\ &= \sum_{gg'} \langle V'|\alpha_{gg'}|V\rangle \langle R'|\Phi_{Fg}\Phi_{F'g'}|R\rangle\end{aligned}$$

$\langle R'|\Phi_{Fg}\Phi_{F'g'}|R\rangle$ is the transition probability for rotation, and $\langle V'|\alpha_{gg'}|V\rangle$ is for vibration.

1. Selection rules for rotation:
   As to $\langle R'|\Phi_{Fg}\Phi_{F'g'}|R\rangle$, we can find a detailed discussion in Ref. [1]. For instance, for the rigid diatomic rotor, the selection rule is $\Delta J = 0, \pm 2$. $\Delta J = -2$ is called the *O* branch, $\Delta J = +2$ the *S* branch and $\Delta J = 0$ the *Q* branch.

   From the rotational level energies, $E_J = BJ(J+1)$ and the selection rules, we get the spectrum as shown in Fig. 8.4.

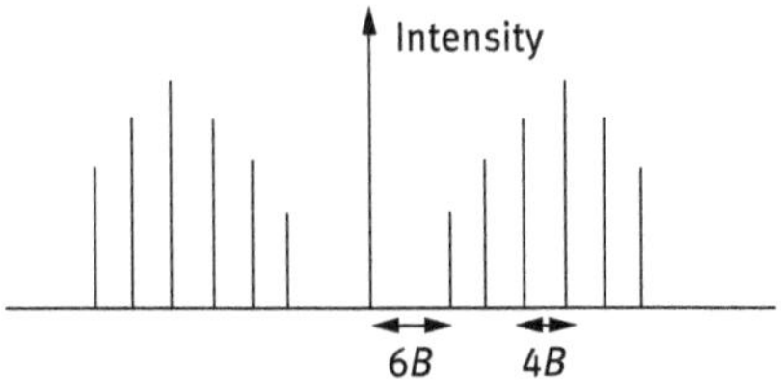

Fig. 8.4: The rotational Raman spectrum.

   For the rotation of symmetric top (Section 2.10), the rule is $\Delta K = 0$, $\Delta J = 0, \pm 1, \pm 2$. But as $K = 0$, $\Delta J = \pm 2$.

2. Selection rules for vibration:
   Since

$$\alpha = \alpha_0 + \sum_k (\partial\alpha/\partial Q_k)_0 Q_k$$

and

$$|I\rangle = |V_1 V_2 \cdots\rangle$$
$$|F\rangle = |V'_1 V'_2 \cdots\rangle$$

then

$$\langle V'|\alpha_{gg'}|V\rangle \sim (\partial\alpha/\partial Q_k)_0 \langle V'_k|Q_k|V_k\rangle \prod_{i \neq k}^{3N-6} \langle V'_i|V_i\rangle$$

The rules are as follows:
1. $(\partial\alpha/\partial Q_k)_0 \neq 0$
2. $\Delta V_k = \pm 1$
3. $\Delta V_l = 0, \ l \neq \mathrm{k}$

This means that there can be only one mode and one vibrational quantum involved in the Raman process.

From the point of view of symmetry, for $\langle F|\alpha|I\rangle$ to be nonzero, the product of Γ_F, Γ_I and Γ_α has to contain a totally symmetric representation.

The symmetries of Γ_α are those of $x^2, y^2, z^2, xy, xz.yz$. See Appendix for character tables.

Note that from the point of view of symmetry, only when α and Q_k are of the same symmetry or belong to the same irreducible representation, $(\partial\alpha/\partial Q_k)_0$ will be nonzero.

For example, for CO_3^{2-} (see Section 5.14)

$$\Gamma_{vib} = A'_1 + A''_2 + 2E'$$

the initial ground state $|I\rangle$ is of total symmetry, then we obtain

$$\Gamma_F\Gamma_\alpha\Gamma_I = \Gamma_F\Gamma_\alpha = \begin{bmatrix} A'_1 \\ A''_2 \\ E' \end{bmatrix} \times \begin{bmatrix} A'_1 \\ E' \\ E'' \end{bmatrix}$$

By reduction, we find that A'_1 and E' modes are Raman allowed and A_2'' mode is not Raman allowed.

For the system possessing inversion center, Γ_α is of g symmetry. For $\Gamma_F\Gamma_\alpha$ to contain the totally symmetric representation A_g, Γ_F has to be of symmetry Γ_g. On the other hand, in IR process, $\Gamma_\sigma(\sigma = x, y, z)$ is of u symmetry; only when Γ_F is of u symmetry, $\Gamma_F\Gamma_\alpha$ will contain A_g. The conclusion is that in a system possessing inversion center, modes that are IR active cannot be Raman active, and vice versa. This is the rule of mutual exclusion.

8.5 Polarizability

From Section 8.1, we have the following:

$$\rho_l = \frac{3\beta^2}{45\bar{\alpha}^2 + 4\beta^2}$$

$$\rho_n = \frac{6\beta^2}{45\bar{\alpha}^2 + 7\beta^2}$$

In the Raman scattering, these formulae still hold, provided that polarizabilities are replaced by the ones that are related to the Raman process.

For the Raman process,

$$\bar{\alpha} = \langle F|\bar{\alpha}|I\rangle$$

Since $\Gamma_{\bar{\alpha}}$ and Γ_I are of total symmetry, if Γ_F (or Q_k) is not of total symmetry, then $\bar{\alpha} = 0$ and

$$\rho_l = \tfrac{3}{4}$$

$$\rho_n = \tfrac{6}{7}$$

attain their maxima.

If Q_k is of spherical symmetry, β=0 and

$$\rho_l = \rho_n = 0$$

attain their minima. For other cases (Q_k is of total symmetry),

$$0 < \rho_l < \frac{3}{4} \qquad 0 < \rho_n < \frac{6}{7}$$

Hence, from ρ_l , ρ_n, we can infer the symmetry of Q_k. Note that the total symmetry is not necessarily spherical symmetry.

A general principle is that the polarizability of bond stretch is larger than that of bending. This is because the electrons in a bond are more affected by the bond stretch than by the bending. Furthermore, the polarizability along a bond is larger than that vertical to it. In addition, when a bond contracts, its polarizability becomes smaller (more bondage by the nuclei). Conversely, when a bond stretches, its polarizability is larger.

Often the reciprocal of polarizability is plotted. This is the polarizability ellipsoid. As an example, for CO_2, the polarizability ellipsoid for the symmetric vibration υ_1 is shown in Fig. 8.5.

If $+Q_1$ and $-Q_1$ are the stretch and contraction of the C–O bond, then the relation of α and Q_1 is as shown in Fig. 8.6. We have

$$(\partial\alpha/\partial Q_1)_0 \neq 0$$

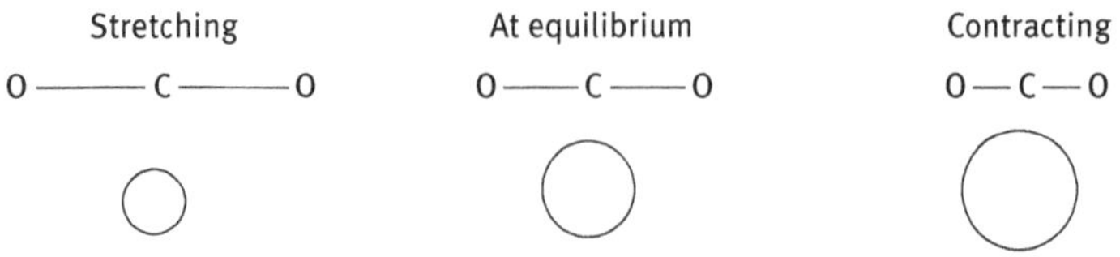

Fig. 8.5: The polarizability ellipsoid for the symmetric vibration υ_1 of CO_2.

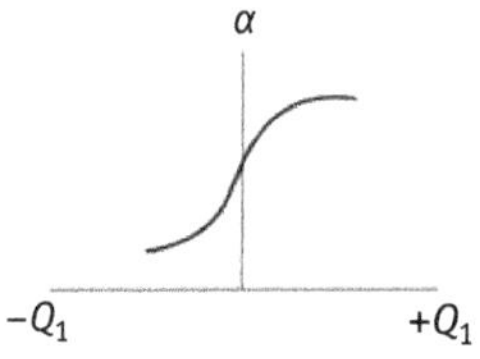

Fig. 8.6: The relation of α and Q_1 coordinate of CO_2.

Hence, υ_1 is Raman allowed. Note that υ_1 is of total symmetry.

Figure 8.7 shows the relation of α and Q_3 of υ_3 mode, which is of the form: $\overleftarrow{O}-\vec{C}-\overleftarrow{O}$

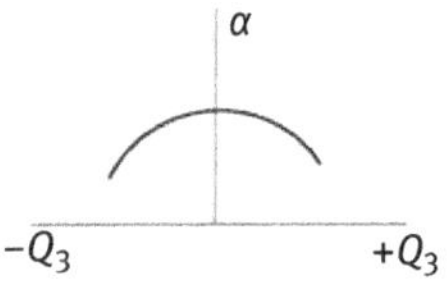

Fig. 8.7: The relation of α and Q_3 of υ_3 mode of CO_2.

Then, $(\partial\alpha/\partial Q_3)_0 = 0$. υ_3 is not Raman allowed. Note that υ_3 is not of total symmetry.

8.6 Wolkenstein's bond polarizability theory

Wolkenstein assumed that molecular polarizability, α, could be expressed as follows:

$$\alpha = \sum_t \alpha_t + \sum_a \alpha_a$$

where α_t and α_a are, respectively, the bond polarizability (in the coordinates with the bond as the principal axis) and the atomic polarizability that is spherical.

For ν_k mode, we have the following expansion in terms of the internal (bond stretch) coordinates S_t:

$$\partial\alpha/\partial Q_k = \sum_t \frac{\partial\alpha}{\partial S_t}\frac{\partial S_t}{\partial Q_k}$$

From the Raman intensity that is related to $(\partial\alpha/\partial Q_k)^2$ and $\partial S_t/\partial Q_k$ that can be obtained by the normal mode analysis, Wolkenstein determined $\partial\alpha/\partial S_t$ ($\approx \partial\alpha_t/\partial S_t$) as shown in Tab. 8.1. From the table, it is seen that these C–H bonds share very similar polarizabilities though they are in different molecules. This means that molecules are built up by chemical bonds just like a room is built up from bricks. The final molecular shapes and properties can be very divergent; however, locally for the same constituents, the properties are quite the same.

Tab. 8.1: C–H bond polarizabilities in different molecules.

Molecule	$\partial\alpha/\partial r_{C-H}(\mathring{A}^2)$	Molecule	$\partial\alpha/\partial r_{C-H}(\mathring{A}^2)$
CH_4	1.04	C_6H_6	1.00
C_2H_6	1.10	C_2H_2	1.02
C_2H_4	1.04		

8.7 Resonance Raman effect

From Section 8.3, we have

$$\left(\alpha_{\rho\sigma}\right)_{ab} = \frac{1}{\hbar}\left\{\sum_{n\neq a}\frac{\langle a_0|\mu_\sigma|n_0\rangle\langle n_0|\mu_\rho|b_0\rangle}{\omega_{an}+\omega} + \sum_{m\neq b}\frac{\langle a_0|\mu_\rho|m_0\rangle\langle m_0|\mu_\sigma|b_0\rangle}{\omega_{bm}-\omega}\right\}$$

It is noted that as $\omega \to \omega_{na}$, $\left(\alpha_{\rho\sigma}\right)_{ab}$ is enhanced significantly and so are the Raman intensities. This is the resonance Raman effect. Then, the Raman excited state can be the eigenstate for which there is longer time for the vibronic coupling to accomplish, hence resulting in larger probability when compared to the nonresonance case. This leads to the enhancement of Raman intensity.

For instance, in $>C=C<$, the $n \to \pi^*$ transition is about 250 nm. If the scattering light is close to the wavelength 250 nm, the Raman signal will be much stronger than that off this wavelength, as shown in Fig. 8.8. We note that the intensity in (b) is larger than that in (a). The exciting wavelengths for (a) and (b) are 632.8 and 337.1 nm, respectively.

Albrecht has studied resonance Raman in detail based on the quantal treatment (Ref. [2]). He pointed out that not all modes are enhanced equally. Instead, only the mode a of symmetry Γ_a that can couple two nearby electronic states $|s\rangle$, $|e\rangle$ of symmetries Γ_s, Γ_e are resonance Raman active in addition to the condition: $\omega \cong \omega_{gs} \cong \omega_{ge}$. That is, $\Gamma_s \times \Gamma_e \supset \Gamma_a$. (Since the electronic density of states is dense

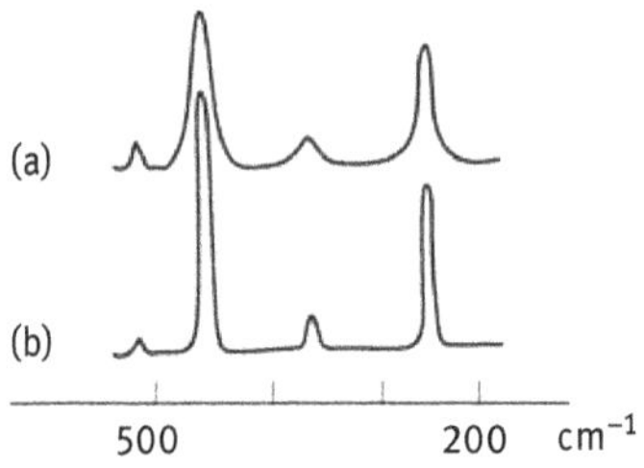

Fig. 8.8: The resonance Raman effect. The exciting wavelengths for (a) and (b) are 632.8 and 337.1 nm, respectively.

for molecules, it is not an issue to have such two nearby states.) Alternatively, by the selection rule, we may easily find out the symmetries of mode a, $|s\rangle$ or $|e\rangle$ if the symmetries of any two of them are known. Indeed, Raman effect provides the information of the coupled electronic states in addition to the vibrational modes.

8.8 Hyper-Raman effect

As the scattering light intensity (with ω_L) is very high, it would lead to nonlinear effect. Then, the induced μ_{ind} is not simply proportional to the electric field ε of light:

$$\mu_{\text{ind}} = \alpha \cdot \varepsilon + \frac{1}{2}\beta\varepsilon \cdot \varepsilon + \frac{1}{6}\gamma\varepsilon\varepsilon\varepsilon$$

where the first term is related to Raman, and the second term is related to the hyper-Raman effect, for which we have

$$\omega_s = 2\omega_L \mp \omega_k$$

Its selection rule is determined by parameter β.

The Raman process is a two-photon process. Other three-photon process, such as stimulated Raman and coherent anti-Stokes Raman scattering, are not treated here.

Exercises

8.1 For the scattering configuration $Y(ZX)Y$, write its scattered light intensity.

8.2 Discuss how to employ the Raman method to analyze the vibrational mode symmetry in liquid, gas and solid.

8.3 Discuss the potential applications of resonance Raman in detecting the materials of trace amount.

8.4 Try to determine the structure of BF_3 based on the following spectral information:

Tab. 8.2: The spectral information of BF_3.

$^{11}BF_3$	$^{10}BF_3$	Raman intensity	IR intensity
480.4 cm^{-1}	482.0 cm^{-1}	Medium	Strong
691.3	719.5	–	Strong
888	888	Strong	–
1,445.9	1,497	–	Very strong
1,831	1,928	–	Weak
2,903.2	3,008.2	–	Weak

8.5 Use simple diagrams to show the polarizability change of the following vibrational modes. Arrows show the atomic displacements.

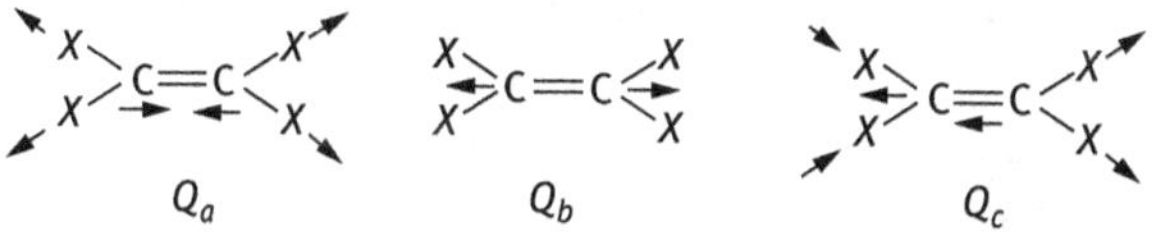

8.6 Try to analyze the vibrational modes of benzene. Which modes are Raman orIR active?

References

[1] Wilson EB, Decius JC, Cross PC. Molecular Vibrations. New York: McGraw-Hill, 1955.
[2] Albrecht AC.J. Chem. Phys. 1961, 34: 1476.

9 Bond polarizability

9.1 Raman intensity

A vibrational spectrum (both i.r. and Raman) has two characteristics: one is the spectral peak position, which is an exposition of the force field in the molecule and the other is the spectral peak intensity, which is an exposition of the delicate information concerning the interaction among the nuclei and electrons. Since the stretching of a bond is more difficult than its bending, or that the force constant of the bond stretch is larger than that of the bend (for instance, in C_2H_4, their ratio for the C–H bond is 5.1:0.3.), the spectral peak due to bond stretch is, in general, in the higher frequency region, while that due to the bend is in the lower frequency region. The molecular constituents, such as C–C, C–H or C=O, may possess similar force constants even in *different* molecules. For instance, the stretch of C=C always appears around $1,700\ \mathrm{cm}^{-1}$ and that of C–H stretch around $3,000\ \mathrm{cm}^{-1}$. Hence, we can expect that various molecules may possess similar spectra if they possess common constituents. This is particularly obvious in the family of organic compounds, since they are mostly composed of C–C, C=C, C=O, C–H and so on.

Molecular normal mode is the motion of a molecule as a whole, including bond stretches and bends, though with various proportions. Normal mode analysis shows that there are the modes that are composed mostly of the bond stretches. This is not difficult to understand since, as mentioned previously, the stretch force constant is one order larger than the bend. The number of this type of modes should be equal to the bond number in a molecule. Though this rule is simple, it is very useful for our interpretation of a spectrum.

Oxalate has five bonds: one C—C and four equivalent C—O bonds. Table 9.1 shows its vibrational assignments [1]. Therein, we see there are five modes that are mainly composed of the stretches of C–O and C–C bonds. Their symmetries are $\upsilon_1(a_{1g})$, $\upsilon_3(a_{1g})$, $\upsilon_4(b_{1g})$, $\upsilon_9(b_{2u})$, $\upsilon_{11}(b_{3u})$. This can be easily confirmed by the group theoretical analysis using the internal coordinates shown in Chapter 5.

The fact that the force constants of the bond stretch and bend are of one order difference in magnitude is a very useful concept. In the normal mode analysis, one can thereby block-diagonalize the force field matrix into submatrices for the convenience of the subsequent analysis. For details, see Reference [2].

For various molecules, their spectra may show similar spectral peaks if they share common constituents. However, their spectral (relative) intensities may vary significantly. This is particularly eminent for the Raman spectra. Raman intensity is related to the molecular polarizability derivative with respect to the normal coordinate. Molecular polarizability is an indication of the confinement of electrons in the molecule. For a kind of bond in different molecules, its polarizability difference is

https://doi.org/10.1515/9783110625097-009

Tab. 9.1: Vibrational assignments of oxalate ion.

D_{2h}			Assignment	Frequency (cm^{-1})
a_{1g}	υ_1	C–O	Stretching	1,488
	υ_2	COO	Deformation	904
	υ_3	C–C	Stretching	445
b_{1g}	υ_4	C–O	Stretching	1,660
	υ_5	COO	In-plane rocking	300
b_{2g}	υ_6	COO	Out-of-plane rocking	1,305
a_{1u}	υ_7		Torsion	–
b_{1u}	υ_8	COO	Out-of-plane rocking	500
b_{2u}	υ_9	C–O	Stretching	1,555
	υ_{10}	COO	In-plane rocking	160
b_{3u}	υ_{11}	C–O	Stretching	1,300
	υ_{12}	COO	Deformation	766

larger than its bond strength. The physics is that in different molecules, the same bond will show very similar strength. However, because of different environments, its electronic polarizability may show significant divergence. We should recognize that the spectral peak *position* only show limited information. A huge amount of information is embedded in the spectral peak *intensity*. This is just analogous to the X-ray crystal diffraction. We know that the diversity of crystal structures is significant. However, their symmetries (which is purely geometric and not related to the physics) are very limited. There are only 230 space groups. As to the diffraction patterns, they are only related to the 32 point groups. The diversity is embedded in the diffracted spot intensities since the Fourier transform of the spot intensities shows the distribution of electrons in the crystal, thereby one can deduce the crystal (molecular) structure. In other words, different crystals can possess the same space group, hence, similar diffraction patterns. However, their spot intensities may be overwhelmingly different. For crystals, we are surely not satisfied with their diffraction patterns. This will only offer limited information. We have to pay attention to the diffracted intensities to explore their structures. Analogously, in spectroscopy, we should not be satisfied with the spectral peak positions. The key is at the analysis of the spectral peak intensities.

9.2 Algorithm for bond polarizability

We now know that the central issue is to retrieve molecular structural information from the Raman mode intensities. In this section, we will show how to elucidate the molecular bond polarizabilities from the Raman mode intensities.

It is known that the Raman mode intensity, I_j, corresponding to the normal mode coordinate Q_j with frequency v_j, can be expressed as

$$I_j \sim I_0(\nu_0 - \nu)^4/\nu_j(\partial\alpha/\partial Q_j)^2$$

where I_0 is the laser exciting intensity, α is the molecular electronic polarizability and ν_0 is the laser frequency.

$(\partial\alpha/\partial Q_j)$ is a quantity of the molecule as a whole. It is crucial to transform it to the parameters of internal coordinates so that the associated physical and chemical contents can be fully appreciated. For this, we know that the relationship between the normal coordinates and the internal or symmetry coordinates $\boldsymbol{S}$ is:

$$\boldsymbol{S} = \boldsymbol{LQ}$$

or

$$\partial S_k/\partial Q_j = L_{kj}$$

Hence,

$$\begin{aligned}\partial\alpha/\partial Q_j &= \sum_k \partial\alpha/\partial S_k \cdot \partial S_k/\partial Q_j \\ &= \sum_k L_{kj}\partial\alpha/\partial S_k = \sum_k (\boldsymbol{L}^{\mathrm{T}})_{jk}\partial\alpha/\partial S_k\end{aligned}$$

Raman intensity I_j is then is ($\boldsymbol{L}^{\mathrm{T}}$ is the transpose of $\boldsymbol{L}$):

$$I_j \sim (\nu_0 - \nu_j)^4/\nu_j\left[\sum_k (\boldsymbol{L}^{\mathrm{T}})_{jk}\partial\alpha/\partial S_k\right]^2$$

Taking the square root of both sides, we have

$$\pm\sqrt{I_j} \sim (\nu_0 - \nu_j)^2/\sqrt{\nu_j}\sum_k (\boldsymbol{L}^{\mathrm{T}})_{jk}\partial\alpha/\partial S_k$$

The sign ± cannot be determined from the experiment. This is the phase problem. In the X-ray structure determination, we are encountered with a similar problem. For Fourier transforming the spot intensities to elucidate the electronic density distribution in the crystal, phases have to be determined first.

The issue here is that from the intensity only the square of $\left[\sum_k (\boldsymbol{L}^{\mathrm{T}})_{jk}\partial\alpha/\partial S_k\right]$ can be known, that is, though the absolute value of $\left[\sum_k (\boldsymbol{L}^{\mathrm{T}})_{jk}\partial\alpha/\partial S_k\right]$ can be known, its true value is unknown.

For convenience, we define

$$a_{jk} = (\nu_0 - \nu_j)^2/\sqrt{\nu_j}(\boldsymbol{L}^{\mathrm{T}})_{jk}$$

Then, we have the relation

$$\begin{bmatrix} P_1\sqrt{I_1} \\ P_2\sqrt{I_2} \\ \vdots \\ P_{3N-6}\sqrt{I_{3N-6}} \end{bmatrix} = \begin{bmatrix} \\ a_{jk} \\ \\ \end{bmatrix} \begin{bmatrix} \partial\alpha/\partial S_1 \\ \partial\alpha/\partial S_2 \\ \vdots \\ \partial\alpha/\partial S_{3N-6} \end{bmatrix}$$

if only *relative* intensities I_j and $(\partial\alpha/\partial S_k)$ are adopted. Here, P_i is + or −. a_{jk} can be calculated via normal mode analysis.

Therefore, if P_i can be determined, then by this equation, $(\partial\alpha/\partial S_k)$ can be obtained. It is the electronic polarizability derivative with respect to the internal or symmetry coordinate. We call it the molecular bond polarizability, or simply, the bond polarizability by which much structural information can be obtained.

In Chapter 8, we have mentioned the Wolkenstein approximation. The approximation supposes that $\partial\alpha/\partial S_k$ is nonzero only for stretch coordinate S_k. By this approximation, the above matrix equation can be reduced to

$$\begin{pmatrix} P_1\sqrt{I_1} \\ \vdots \\ P_w\sqrt{I_w} \end{pmatrix} = \begin{pmatrix} d_{ij} \end{pmatrix} \begin{pmatrix} \partial\alpha/\partial S_1 \\ \vdots \\ \partial\alpha/\partial S_w \end{pmatrix}$$

with w being the number of the bond stretch coordinates or just the bond number in the molecule.

In fact, if there is definite symmetry in the molecule, then only totally symmetric coordinates S_k are required. This is because of the fact that:

1. If S_k is totally symmetric, then $S_k = \frac{1}{\sqrt{N}}\sum_i^N r_i$
   with r_i being the bond displacement. Hence, $\partial\alpha/\partial S_k \approx \partial\alpha/\partial r_i$.
2. For those S_k not of total symmetry, the corresponding $\partial\alpha/\partial S_k$ will be much smaller. This is due to the fact that S_k is the linear combination of bond stretch coordinates with opposite signs.

After these reductions, we have

$$\begin{pmatrix} \partial\alpha/\partial S_1 \\ \vdots \\ \partial\alpha/\partial S_m \end{pmatrix} = \begin{pmatrix} d_{ij} \end{pmatrix}^{-1} \begin{pmatrix} P_1\sqrt{I_1} \\ \vdots \\ P_m\sqrt{I_m} \end{pmatrix}$$

where $S_1 \cdots S_m$ are of total symmetry and $I_1 \cdots I_m$ are those peak intensities that are mostly due to the bond stretches. (Of course, we may consider those bend coordinates that are closely coupled to the stretch coordinates. Then, we obtain the bond polarizabilities of both stretch and bend coordinates.)

It needs to be emphasized that this algorithm is based on the analysis of a set of Raman intensities, instead of an individual one. If due to certain reasons that there is uncertainty in a specific peak intensity, then the result by this algorithm may not be so seriously affected as if the analysis is based on that individual peak. In this sense, this algorithm is more credible than that based on an individual peak.

The determination of phases $\{P_i\}$ is more difficult. For example, if there are 10 bond polarizabilities to elucidate, 10 spectral intensities and hence 10 P_i's are needed. Each P_i can be either + or −. Hence, there are 1,024 possible combinations of phases or 1,024 possible sets of bond polarizabilities. The central issue then is to nail down the unique one. At a first glance, this seems a formidable task. However, the situation is not so pessimistic. The point is that among the numerous solution sets, very few can satisfy the realistic physical requirements. In certain situations, the allowed set is even unique. What are the physical requirements? For instance, the bond stretch polarizability is always or in most situations larger than that of the bend; the polarizability of a double bond is larger than that of a single bond; the bond polarizabilities of C–C and C–H are roughly of the same order in magnitude and the same positive sign. (This is confirmed by theoretical calculation.) Besides, sometimes, certain peak intensities are much smaller. Their corresponding phase signs, no matter "+"or "−", offer very similar results. This also reduces the indeterminate choices.

In this algorithm, normal mode analysis is required to obtain $[d_{ij}]$ matrix. For the analysis, we need to know the molecular structure, in particular, the bond lengths, angles and force constants. These parameters are usually available in the literature. Meanwhile, quantum chemical calculations by GAUSSIAN software can be employed to offer valuable data. In practice, we may first use those force constant data either from the literature or by calculation as initial inputs to obtain the calculated mode frequencies, then by least fitting to the experimental mode frequencies to obtain the optimum force constants and L_{kj}.

We need to stress only that relative intensities and hence relative bond polarizabilities are of our concern. This greatly reduces the inaccuracies due to the instrumentation. These relative quantities are enough to serve our purpose. For more accurate measurement of the Raman intensity, the effects by the grating mirror due to its differential reflectivity in different frequency domains and polarizations, and by the detector due to its differential responses in different frequency domains, have to be considered and corrected, if needed. For the modern state-of-the-art instrumentation, these corrections are becoming unnecessary.

9.3 Surface-enhanced Raman intensity

Above formalism can be applied to systems of interest, irrespective of if they are gaseous, liquid or solid phase. For a system, its relative Raman intensities are in general rather fixed, provided that its structure is not seriously changed under the environmental variation, say, temperature. In 1974, a new phenomenon, called the surface-enhanced Raman scattering (SERS), was found, which shows that the Raman cross section of nitrogen-containing molecules when adsorbed on the roughened silver and gold surfaces (sol or electrode) could be augmented up to 10^6-fold. The most prominent phenomenon is that for these molecules, their relative intensities are dependent on the applied electrode voltage. The situation is that this is not due to the abundance of the adsorbed species. Instead, it is due to the changing of the intrinsic property of the adsorbed species. Figure 9.1 shows the *relative* spectral intensity variation of piperidine adsorbed on the silver electrode under various applied voltages (with respect to the standard Calomel electrode, SCE [3]). We note that the relative Raman intensities of its solution are also different from the SERS case.

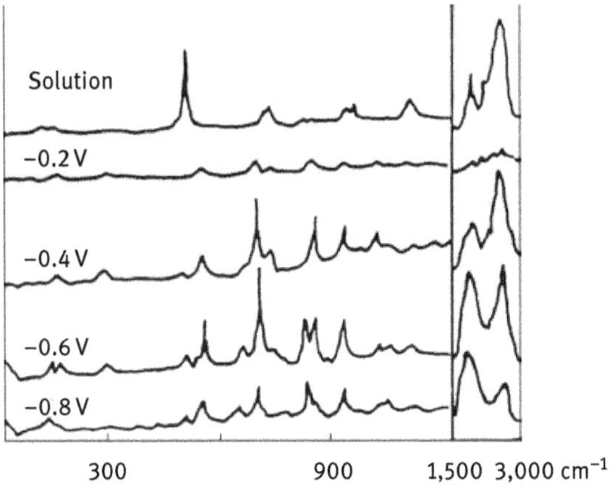

Fig. 9.1: The variation of Raman intensities of piperidine on the silver electrode under various applied voltages (SCE) and in solution.

Up to this time, the very SERS mechanism has not yet been nailed down. It is generally recognized that SERS is related to the charge transfer between the adsorbates and roughened surface. This is the so-called charge transfer or chemical effect. Another possible mechanism is due to the exceedingly strong electric and magnetic (EM) field around the tip-like structure, where the adsorbates sit on the roughened surface. This is called the EM or physical-mechanism.

The issue now is: if we can obtain the bond polarizabilities from the SER intensities, we will obtain the properties of the adsorbates and even the SERS mechanism!

9.4 Bond polarizabilities from SER intensities

The molecular structure of piperidine and its bond stretch coordinates are shown in Fig. 9.2. Its symmetry group is C_S. We need to calculate the bond polarizabilities of its 10 bond stretches. Hence, 10 SER intensities are required whose normal coordinates are composed mostly of bond stretches (C–C and C–H).

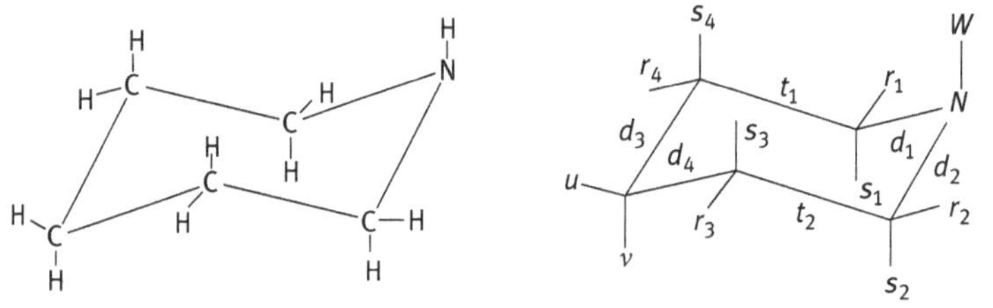

Fig. 9.2: The structure and bond stretch coordinates of piperidine.

The symmetry coordinates adopted are:

$$S_1 = r_1 + r_2$$

$$S_2 = r_3 + r_4$$

$$S_3 = s_1 + s_2$$

$$S_4 = s_3 + s_4$$

$$S_5 = d_1 + d_2$$

$$S_6 = d_3 + d_4$$

$$S_7 = t_1 + t_2$$

$$S_8 = u$$

$$S_9 = v$$

$$S_{10} = w$$

$\partial\alpha/\partial r_{\text{C-C}}$ and $\partial\alpha/\partial r_{\text{CH}}$ are both positive. This offers us a clue to solve the phase problem. We will choose an arbitrary set of $\{P_1 \cdots P_{10}\}$, and calculate $\{\partial\alpha/\partial S_1 \cdots \partial\alpha/\partial S_{10}\}$ by the matrix equation in Section 9.2. The judgment for the accepted phases is that all $\partial\alpha/\partial r_{\text{C-C}}$, $\partial\alpha/\partial r_{\text{CH}}$ are positive when $\partial\alpha/\partial u$ is normalized to 1. While for $\partial\alpha/\partial r_{\text{CN}}$ and $\partial\alpha/\partial r_{\text{NH}}$, no requirements will be imposed.

The results are:

1. For -0.2 V, among the 10 I_i, there are four zero intensities. The numbering of the intensities can be arbitrary. However, it should be consistent with the

ordering in $\left[d_{ij}\right]^{-1}$. For a clear presentation, we will not show the numbering explicitly. This will not affect our demonstration. Hence, there are 64 possible phase sets. However, only one survives under the requirements.
2. For -0.6 V, there are 32 possible phase sets. Only one survives, which is identical to that in the previous case.
3. For -0.4 V, there are 32 phase sets. Among them, two survive. The difference lies on one phase. Since its corresponding intensity is very small, the variation of this phase will not offer much different results. One of the two accepted phase sets is identical to that in the previous two cases.
4. For -0.8 V, among the 32 possible sets, eight sets survive. However, there is one that is identical to that in the previous cases. This set also offers more reasonable $\partial\alpha/\partial S_i$ as $\partial\alpha/\partial u=1$.

Thus, the phase problem is concluded. The elucidated bond polarizabilities are given in Tab. 9.2.

Tab. 9.2: The elucidated bond polarizabilities of piperidine with $\partial\alpha/\partial u=1$ at -0.4 V.

Bond polarizability	**Voltage (V_{SCE})**			
	−0.2	**−0.4**	**−0.6**	**−0.8**
$\partial\alpha/\partial u$	0.8	1.0	1.0	1.5
$\partial\alpha/\partial v$	0.3	0.2	0.2	1.2
$\partial\alpha/d_3$	5.1	6.6	5.1	19.0
$\partial\alpha/\partial t_1$	1.2	1.6	1.3	4.9
$\partial\alpha/\partial r_1$	0.3	~0.0	0.3	0.5
$\partial\alpha/\partial r_3$	0.4	0.2	0.4	0.8
$\partial\alpha/\partial s_1$	~0.0	0.1	~0.0	0.2
$\partial\alpha/\partial s_3$	0.3	0.2	0.2	0.5
$\partial\alpha/\partial d_1$	0.7	0.9	0.9	3.2
$\partial\alpha/\partial w$	~0.0	~0.0	~0.1	0.1

From the results, we may infer the following:
1. As the voltage is shifted from -0.2 to -0.8 V, the bond polarizabilities increase in general. However, their tendencies are different. This shows that as the voltage is more negative, the adsorption is weaker and the bond polarizabilities is larger due to less bondage on the adsorbates from the electrode surface.
2. The bond polarizabilities on the carbon ring are much larger than those of the C–H bonds. Both C–C and C–H are single bonds. Their polarizabilities should be roughly equal. This strongly suggests that the electronic/charge abundance along the ring skeleton is exceedingly enhanced and this is very probably due to the electronic sharing between the adsorbed piperidine molecule and the

electrode surface. This is just the chemical effect. As the voltage is shifted from −0.4 to −0.6 V, the polarizabilities along the ring are smaller while the others are smaller when the voltage is between −0.2 and −0.4 V. This difference shows the very delicate behaviors of the adsorbates in SERS.

3. We note that

$$\partial\alpha/\partial d_3 > \partial\alpha/\partial t_1 > \partial\alpha/\partial d_1$$
$$\partial\alpha/\partial v \geq \partial\alpha/\partial s_3 > \partial\alpha/\partial s_1$$

and

$$\partial\alpha/\partial u > \partial\alpha/\partial r_3 > \partial\alpha/\partial r_1$$

This shows that the adsorption point is on the N atom. As a bond is farther away from the N atom, its polarizability will be larger due to less bondage from the adsorption site.

4. $\partial\alpha/\partial w(\partial\alpha/\partial r_{N-H})$ is very small, showing that electrons around the $N-H$ bond are very scarce due to the charge transfer away to the electrode surface.
5. We note that

$$\partial\alpha/\partial r_3 \geq \partial\alpha/\partial s_3,$$
$$\partial\alpha/\partial u > \partial\alpha/\partial v,$$
$$\partial\alpha/\partial r_1 \geq \partial\alpha/\partial s_1,$$

In other words, the equatorial C–H possesses larger polarizability than the axial one does. (At −0.4 V, $\partial\alpha/\partial r_1$ and $\partial\alpha/\partial s_1$ are very small and should not be considered as a counter-case.) This can be attributed to the electric field vertical to the adsorption surface. As a piperidine molecule is adsorbed on the surface via its N atom, its equatorial C–H bonds are more or less parallel to the electric field. Hence, their electronic behavior is more disturbed, leading to their larger polarizabilities. While the axial C–H bonds are more or less vertical to the electric field and, therefore, possess smaller polarizabilities. This is consistent with the prediction by the EM mechanism.

In conclusion, we have

1. From the bond polarizabilities, it can be inferred that piperidine molecule is adsorbed vertically on the surface with the adsorption site at the N atom.
2. Both the chemical and physical mechanisms are operative as demonstrated by the bond polarizabilities.
3. Table 9.2 shows very delicate behavior of the bond polarizabilities under SERS. A complete SERS model should be able to reproduce these data. In other words, the bond polarizabilities that are derived from the Raman mode intensities set a criterion for the SERS model.

As another example, we show the pyridazine case [4]. The structure and the bond labels of pyridazine are shown in Fig. 9.3.

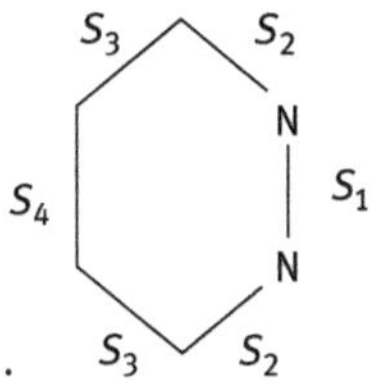

Fig. 9.3: The structure and the bond labels of pyridazine.

As the piperidine case, we can obtain the bond polarizabilities of pyridazine from its SER intensities as listed in Tab. 9.3.

Tab. 9.3: The bond polarizabilities of pyridazine under various voltages.

Polarizability	**Applied voltage (V_{SCE})**		
	−0.4	**−0.6**	**−0.8**
$\partial\alpha/\partial S_1$	3.8	2.3	2.2
$\partial\alpha/\partial S_2$	−1.8	−3.1	−0.4
$\partial\alpha/\partial S_3$	8.9	7.3	7.1
$\partial\alpha/\partial S_4$	7.0	5.7	10.0

We obtain the following inferences from Tab. 9.3:

1. As the voltage is close to −0.8 V, except S_4, all bond polarizabilities become smaller. At this voltage, the polarizability difference among the bonds is also larger. Then the adsorption is stronger. $\partial\alpha/\partial S_2$ attains its maximum at −0.6 V, while for $\partial\alpha/\partial S_4$, it is at −0.8 V. Different bonds behave differently. SERS process is indeed quite complicated.
2. As the voltage shifts from −0.4 to −0.6 V, $\partial\alpha/\partial S_1, \partial\alpha/\partial S_3$ vary more, while the voltage shifts from −0.6 to −0.8 V, $\partial\alpha/\partial S_2, \partial\alpha/\partial S_4$ vary more. This can be interpreted in the following way. As the voltage shifts from −0.4 to −0.6 V, the electrons on the bonding orbital $n_1 + n_2$ formed from the lone-paired orbitals n_1, n_2 of the two N atoms flow to the electrode. Since the electronic density of $n_1 + n_2$ concentrates more on the N–N bond, the charge transfer will make $\partial\alpha/\partial S_1$ smaller. Meanwhile, the electronic shift based on conjugation in the ring will also cause $\partial\alpha/\partial S_3$ smaller. As the voltage shifts from −0.6 to −0.8 V, the charge transfer is from the antibonding orbital $n_1 - n_2$ to the electrode. Since the electronic density of $n_1 - n_2$ is more on the two C–N bonds, the charge

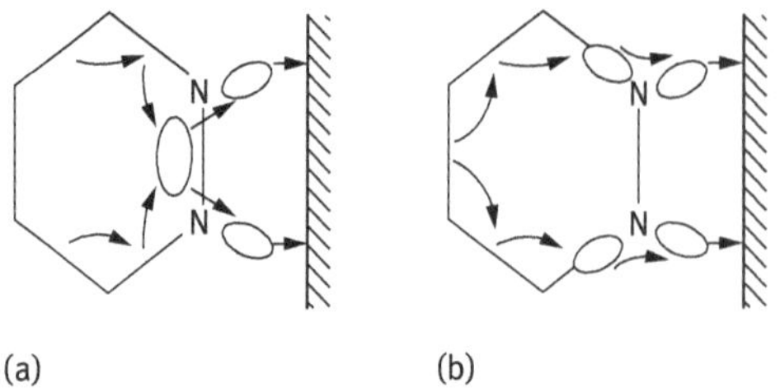

Fig. 9.4: The charge transfer of (a) bonding orbital $n_1 + n_2$ and (b) antibonding orbital $n_1 - n_2$ of the pyridazine molecule to the electrode surface.

transfer will cause $\partial\alpha/\partial S_2$ to vary/decrease more. Meanwhile, by conjugation effect, $\partial\alpha/\partial S_4$ also varies more. These processes are depicted in Fig. 9.4.

3. From the charge transfer model proposed earlier, it is inferred that the frequencies of those modes involving the N–N stretch will decrease, as the voltage shifts from −0.4 to −0.6 V; since then the electronic density on the N–N bonding orbital decreases, and hence, their force constants. When the voltage shifts from −0.6 to −0.8 V, the frequencies of these modes will increase, since then the electronic density on the N–N antibonding decreases, and hence, their force constants become larger. As shown in Tab. 9.4, υ_1, υ_{8a}, υ_{19b}, υ_{12} that involve the N–N bond do show such a tendency under the applied voltages.

Tab. 9.4: The wave number of υ_1, υ_{8a}, υ_{19b}, υ_{12} that involve the N–N bond under the various potentials.

	Potential (V_{SCE})		
Mode	**−0.4**	**−0.6**	**−0.8**
υ_1	976.5	974.5	976
υ_{8a}	1,575	1,573	1,582
υ_{19b}	1,454	1,451	1,453
υ_{12}	1,055	1,052	1,052

In this chapter, we showed how to obtain bond polarizabilities from Raman mode intensities. We also demonstrated the steps to realize a physical idea from its very beginning. It is important not only to obtain a good spectrum but also to interpret its physical/chemical contents. Finally, we note that by this algorithm, the Raman process is closely related to the electronic behavior in a molecule.

References

[1] Ito K, Bernstein HJ. Can. J. of Chemistry. 1956, 34: 170.
[2] Wilson EB, Decius JC, Cross PC. Molecular Vibrations. New York: McGraw-Hill, 1955, Chapter 9.
[3] Tian B, Wu G, Liu G.J. Chem. Phys. 1987,87: 7300.
[4] Wu G.J. Mol. Struct. 1990, 238: 79.

10 Electronic structure of Raman virtual state

10.1 Raman intensity in the temporal domain

We generally refer Raman intensity for a vibrational mode to the area under the spectral contour in the frequency/wave number domain. The spectral intensity in the frequency domain, $I_j'(\nu)$, can be Fourier transformed to the temporal domain, $I_j(t)$. These two intensities are related by $\int I_j'(\nu)e^{i2\pi\nu t}d\nu = I_j(t)$. We note that for $t = 0$; this transformation reduces to $\int I_j'(\nu)d\nu = I_j(t=0) \equiv I_{j0}$. Hence, the integrated Raman intensity over ν is just the temporal intensity at $t = 0$.

10.2 Raman virtual state

The Raman effect is an inelastic two-photon process, in which the scattering photon is first absorbed by a molecule. The photon-perturbed molecule then relaxes and emits a secondary photon instantaneously. The photon-perturbed molecule, in general, may not be in its excited eigenstates. In such a case, we call it the nonresonant Raman process and the photon-perturbed molecule including its relaxation, the Raman excited virtual state or simply the Raman virtual state. The Raman virtual state is just like a wave packet.

It is a common day-to-day experience that in a pan full of water, there can be standing waves that are shaped like the pan. When a pebble is thrown into the pan, the water inside will splash above or even out of the pan in a random way. Similarly, when a molecule (the pan) absorbs a light quantum (the pebble), its electronic distribution (the water) will be disturbed. The disturbed (excited) electronic distribution in such an excitation (the splashing water) is, in general, not stationary or nonresonant and does not correspond to an eigenstate (the standing wave). We know that eigenstates are governed by the molecular nuclei and can be accurately predicted by the Schrodinger equation (just as that the patterns of the standing waves are defined by the pan boundary) while the nonstationary excitation, including its relaxation, that is, the virtual state, is not well defined by the nuclei and is difficult to figure out, though not impossible. This is the physical concept of the Raman virtual state. We cannot simply say that splashing water is not a physical reality because it is not a standing wave; we cannot say that the Raman virtual state is not a real physical entity because it is not an eigenstate.

In the quantum mechanical language, the Raman virtual state can be considered as the superposition/sum of all possible allowed intermediate excited processes/ states as shown in Fig. 10.1.

We demonstrate in this chapter that the electronic information of the virtual state is embedded in the Raman intensity [1].

https://doi.org/10.1515/9783110625097-010

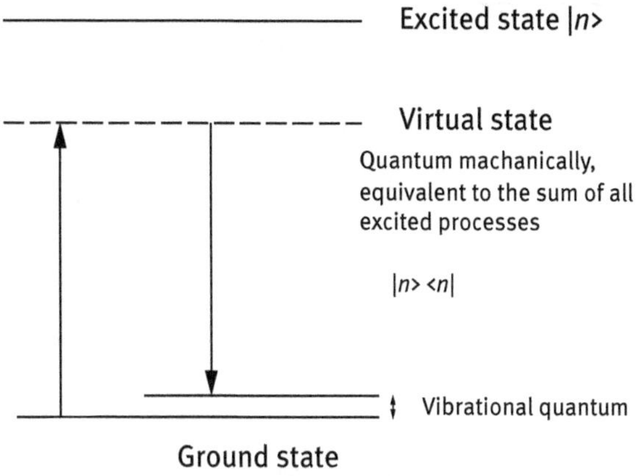

Fig. 10.1: The Raman excited virtual state is equivalent to the sum of all the excited processes, quantum mechanically.

10.3 Raman virtual electronic structure of 2-aminopyridine

We demonstrate the Raman virtual electronic structure of 2-aminopyridine at 514.5 and 632.8 nm excitations as shown in Fig. 10.2. This molecule possesses 13 bonds. We need 13 Raman mode intensities, for which the mode coordinates are mostly of bond stretch to elucidate the polarizabilities of bond (stretch).

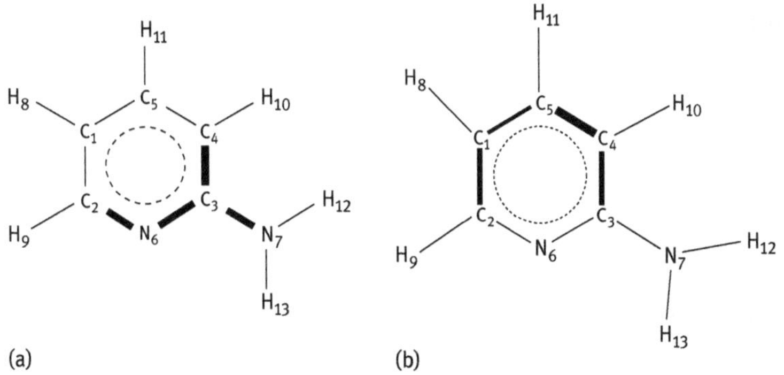

Fig. 10.2: (a) The bond polarizabilities of the excited virtual state of 2-aminopyridine and (b) the bond electronic densities of its ground state The thicker bond segment shows that the bond possesses a more significant polarizability or electronic density.

Figure 10.3 shows its Raman spectra at 632.8 and 514.5 nm excitations. * shows those 13 modes whose intensities are employed for the elucidation of the polarizabilities of bond stretch. Their experimental and fitted Raman shifts together with

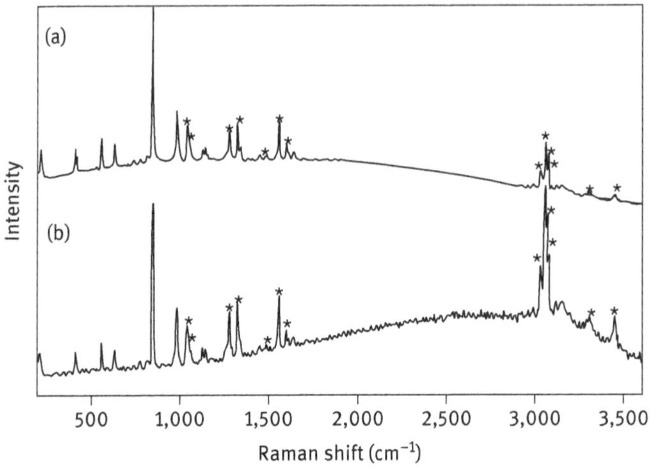

Fig. 10.3: The Raman spectra of 2-aminopyrimidine at (a) 632.8 nm and (b) 514.5 nm excitations. No significant variation of the peak positions is observed by these two excitations. However, there are variations in their intensities as listed in Tab. 10.1. * shows those modes whose intensities are employed for the elucidation of the polarizabilities of bond stretch.

potential energy distributions and relative intensities at 514.5 and 632.8 nm excitations are listed in Tab. 10.1 in which the mode intensities at 1,558 cm^{-1} are normalized to 100 for convenience.

The criteria for the phase choice are (1) the bond polarizabilities of the same kind bonds are of the same sign and (2) the relative magnitudes of C–N and C–C polarizabilities are within 1.5 fold. We found that this value is not rigid. Its variation does not affect the phase solution seriously. Then we obtain only one phase set.

Figure 10.4 shows the 13 bond polarizabilities at 514.5 and 632.8 nm excitations. The bond polarizabilities of C3–N7 are the largest in both 514.5 and 632.8 nm excitations. However, the quantal calculation shows that for the ground state, the electronic density of C3–N7 bond is the least. Figure 10.2 shows this contrast. The implication is that the charges of the virtual state do spread to the outer portion of the molecule so that the bond polarizabilities of outer C–H and N–H bonds enhance more. We note a very minor difference at 514.5 and 632.8 nm excitations: at 514.5 nm excitation, the bond polarizabilities of outer C–H and N–H bonds enhance more, while those on the ring skeleton are depressed more. This implies that 514.5 nm excitation is more powerful that it can excite more charges toward the molecular periphery, resulting in more enhancements in the bond polarizabilities of the outer C–H and N–H bonds. This observation is also presented in Tab. 10.1 that mode intensities with more C–H and N–H stretches enhance more while those mode intensities with more skeletal stretches enhance less at 514.5 nm than at 632.8 nm.

Tab. 10.1: The experimental and fitted Raman shifts together with their potential energy distributions and relative intensities at 514.5 and 632.8 nm excitations. The intensities at 1,558 cm^{-1} by both excitations are normalized to 100 for convenience. Only those distributions in stretches are listed. These are the 13 modes that possess larger portion in bond stretch motion and are employed for the elucidation of the bond polarizabilities – as: antisymmetric, s: symmetric.

Raman Shift (cm^{-1})		**Intensity**		**Potential energy distribution**
Exp.	**Fitted**	**514.5 nm**	**632.8 nm**	
3,447	3,433	109.4	33.4	ν(N7H12, N7H13)$_{as}$ (100)
3,303	3,327	79.4	30.0	ν(N7H12, N7H13)$_{s}$(100)
3,076	3,079	125.3	51.8	ν(C1H8)(84.8),ν(C5H11)(9.8),ν(C2H9) (2.3) ,ν(C4H10) (2.2)
3,060	3,060	300.7	146.9	ν(C5H11)(89.2),ν(*C1H8*)(9.2)
3,045	3,045	41.1	18.8	ν(C4H10)(95.8),ν(C1H8) (2.2),ν(C2H9) (1.1)
3,031	3,025	122.6	49.7	ν(C2H9)(95.3), ν(C1H8) (2.2)
1,599	1,593	24.4	25.4	ν(C2N6)(17.4),ν(C4C5)(17.1),ν(C1C2)(16.1),ν(C3C4) (12.7), ν(C3N7)(1.8)
1,558	1,571	100	100	ν(C3N6) (23.5),ν(C1C5) (20.9), ν(C3C4) (9.5), ν(C4C5) (7.3), ν(C1C2)(4.3),ν(C2N6)(2.2)
1,483	1,485	10.1	14.5	ν(C3N6) (14.4),ν(C3N7)(13.6), ν(C1C5)(5.2), ν(C3C4)(5.0), ν(C1C2)(3.8), ν(C2N6)(3.4), ν(C4C5)(1.6)
1,325	1,319	76.1	73.9	ν(C3N7)(41.8), ν(C2N6)(17.6), ν(C4C5)(3.5), ν(C1C5)(2.0), ν(C3C4)(1.0)
1,280	1,285	102.8	86.6	ν(C4C5)(16.2), ν(C2N6)(16.2), ν(C3N6)(15.7), ν(C1C2)(14.7), ν(C1C5)(6.8), ν(C3N7)(2.2)
1,051	1,051	80.9	101.1	ν(C1C2)(16.1), ν(C3N6)(12.7), ν(C3C4)(8.4), ν(C2N6)(5.3), ν(C1C5)(5.1), ν(C4C5)(4.7)
1,042	1,026	59.5	73.1	ν(C1C5)(40.9), ν(C1C2)(21.5), ν(C4C5)(6.9), ν(C3C4)(3.0), ν(C3N6)(1.8)

We can Fourier transform the spectrum to the temporal domain and obtain the bond polarizabilities as the functions of time during the Raman process. In most cases, the bond polarizabilities are of the exponentially decaying functional form. Around 5 ps (1 ps = 10^{-12} s), the process reaches its full relaxation as shown in Fig. 10.5. The result shows that the bond electronic densities of the ground state calculated by the upper eight occupied levels are close to the bond polarizabilities

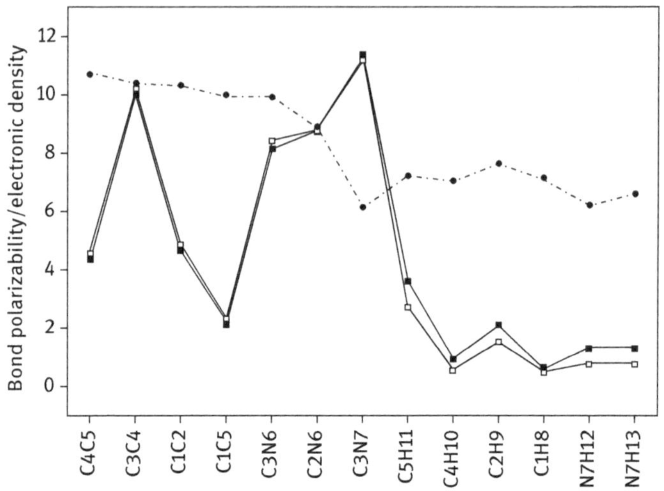

Fig. 10.4: The plots of the bond polarizabilities at 514.5 and 632.8 nm excitations together with the bond electronic densities calculated by RHF/6-31G*. For the convenience of inference, the values of C3–C4 bond are normalized to 10. The bond polarizabilities: (■) for 514.5 nm, (□) for 632.8 nm and (●) for bond electronic densities.

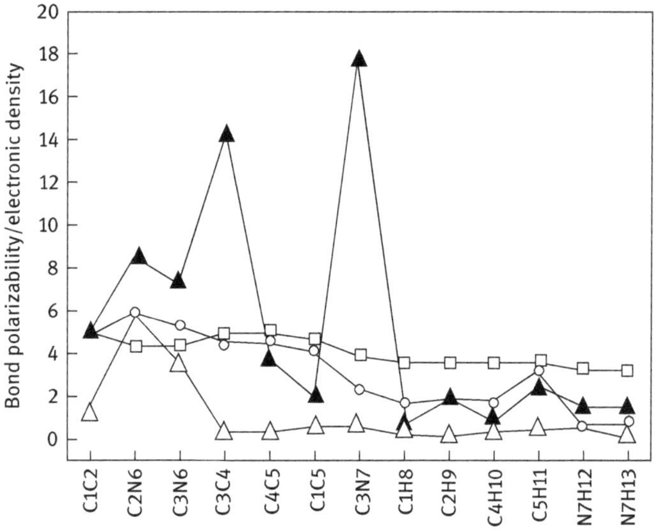

Fig. 10.5: The bond polarizabilities and the calculated bond electronic densities for 2-aminopyriding. ▲ and Δ are for the bond polarizabilities at $t = 0$ and 5 ps. □ and ○ are for the calculated bond electronic densities of the ground state by all the occupied MOs and by only the upper eight MOs, respectively. All the values are relative to the value of C1–C2 bond normalized to 5 for convenience. Note that the profiles by Δ are closer to those by ○. Also note that there is no comparability in values between the bond polarizabilities and the bond electronic densities.

at full relaxation. This means that not all electrons in this molecule are involved in the Raman process. We have studied other molecular cases. However, more cases imply that all electrons are involved in this process. The reason can be that for this case, the upper eight occupied levels are quite separated from the rest of the lower levels so that the electrons in these upper eight levels are more involved in the Raman excitation.

The bond polarizabilities at full relaxation that are close to the bond electronic densities of the ground state by quantal calculation deserve attention. We note that bond electronic density is calculated theoretically, while the bond polarizability is derived from the experimental quantity, that is, the Raman intensity. The consistency of these two quantities is not accidental. It means that our method of obtaining bond polarizability is correct and practical. The significant point here is that the information about the bond electronic densities of the ground state is embedded in the Raman mode intensities.

10.4 Uncertainty relation in Raman virtual state

We have studied the case of ethylene thiourea. It is noted that the relaxation of its temporal bond polarizabilities are in the range about 8 and 5 ps, respectively, at 514.5 and 325 nm excitations. These time durations show roughly the lifetimes of Raman virtual states at 514.5 and 325 nm excitations. This ratio of 8:5 is roughly proportional to the wavelength ratio of the excitations, 5:3. This is in agreement with Heisenberg's uncertainty principle that lifetime is inversely proportional to the excitation energy, or is proportional to the wavelength of the exciting light. This confirms that Raman excited state is not a stationary eigenstate. Hence, it is appropriate to call this state nonstationary or *virtual*.

10.5 Charge distribution in Raman virtual state

The charge redistribution in the virtual state as evidenced by the bond polarizabilities is an interesting topic. The issue is the determination of the proportional constant t between the electron charge e and the bond polarizability $(\partial\alpha/\partial R_k)$: $e = t(\partial\alpha/\partial R_k)$. Since the bond polarizabilities we obtain are of relative magnitudes, this is an impossible issue to solve except that we can obtain absolute Raman intensities, which will not be an easy task, or that we can find another quantity that can impose an extra condition. It will be demonstrated in Chapter 12 that Raman optical activity spectrum can be of such a condition. Before the final determination of the parameter t, in this section, we will formulate the related expressions. We will treat the issue in a classical way. This will give us a concrete idea about the electronic structure of the Raman virtual state.

We first partition each bond polarizability (since charges are concentrated on the bonds, we will only employ the bond polarizabilities of the stretching coordinates in this treatment) evenly to the two atoms forming the bond. Then, all the polarizabilities on an atom are summed, scaled and combined with its net charges in the ground state.

Here, we will pay more attention to the charges on the atoms. When a molecule is in its ground state, there are charge distributions between the atoms. This is the chemical bond. For convenience, one usually attributes the charges on a bond to the atoms that form the bond. This is the Mulliken charge (e_j^0, subscript j stands for the atomic numbering). Thus, it is an *effective* charge. Since a molecule is neutral, the sum of all the Mulliken charges is zero, that is, $\sum_j e_j^0 = 0$. In the Raman virtual state, the sum of the net charges on all the atoms is still zero. This means that the charges on the atoms that are derived from the bond polarizabilities after scaling as shown previously have to come from the Mulliken charges in the ground state. We know that close to the complete relaxation of the Raman excitation, the bond polarizabilities are parallel to the bond electronic densities in the ground state, which are calculated by the entire occupied MO, that is, all the electrons in a molecule. (At least, in the case we will deal with.) This implication is that all the electrons in a molecule contribute to the excited charges in the Raman virtual state with equal probability. The electron number on the jth atom in the ground state is $z_j = Z_j - e_j^0$, with Z_j being the atomic number. Hence, in the virtual state, the excited charges contributed from the jth atom will be as follows:

$$t \bullet z_j \bullet \sum_l \partial\alpha/\partial R_l/N, \ (t > 0)$$

where N is the total electron number in a molecule, that is, $N = \sum_j Z_j = \sum_j z_j$. $\sum_l \partial\alpha/\partial R_l$ is the sum of all the bond polarizabilities. Since the distributions of the excited charges on the various bonds (as evidenced by the bond polarizabilities) are different and if we partition each bond polarizability evenly to the two atoms forming the bond, then the effective charge on the jth atom is as follows:

$$e_j = e_j^0 + t \bullet z_j \bullet \sum_l \partial\alpha/\partial R_l/N - t\sum_{k_j} \partial\alpha/\partial R_{k_j,j}/2$$

where k_j is the index for the bonds that are associated with the jth atom. We note that $\sum_j z_j \sum_l \partial\alpha/\partial R_l/N = \sum_j \sum_{k_j} \partial\alpha/\partial R_{k_j,j}/2 = \sum_l \partial\alpha/\partial R_l$ and $\sum_j e_j^0 = 0$, hence $\sum_j e_j = 0$. This satisfies the condition that the net charge of the Raman virtual state is zero. We note that if

$$z_j \sum_j \partial\alpha/\partial R_l/N - \sum_{k_j} \partial\alpha/\partial R_{k_j,j}/2$$

is negative, then the jth atom will acquire electrons in the Raman excitation; otherwise, it will lose electrons.

As with t, if we scale the bond polarizabilities to the unit of the elementary charge, then we have, $t\sum_l \partial\alpha/\partial R_l \le N$ or $t \le N/\sum_l \partial\alpha/\partial R_l$.

The determination of t will be finalized in Section 12.5 of Chapter 12.

10.6 Remarks

The bond polarizabilities retrieved from Raman mode intensities reflect the charge distribution (electronic structure) in the virtual state, which is through the mechanism of vibronic coupling. The bond polarizabilities we obtained previously are from the intensities of Stokes line. How about the bond polarizabilities derived from the anti-Stokes intensities? Does the difference of the two processes result in the variation of bond polarizabilities or the electronic structure of the virtual state? This is an interesting topic. The experiment for taking the anti-Stokes intensities is not easy since the population on the $n = 1$ vibrational state is very small. The anti-Stokes intensities, in general, are very small.

Reference

[1] Wu G. Raman spectroscopy. Singapore: World Scientific, 2017, Chapter 4.

11 Circular dichroism

11.1 Introduction

Chiral molecules are those molecules whose structures cannot be superimposed on their mirror images. Organic molecules possessing tetrahedral carbon atom in which the four bonds are connected with different moieties are typical chiral molecules.

The polarization of a linearly polarized light will be rotated as it passes through a chiral molecule. The right- and left-hand enantiomers will be rotated as it possess just the opposite rotation power. This activity is related to the intrinsic property of the chiral molecule. This phenomenon also happens in the absorption and Raman scattering; in these the chiral molecule possesses different absorption and scattering power for the right and left circularly polarized light. The difference is usually very small, only 10^{-3} to 10^{-4}. The optical activity in IR, visible and ultraviolet absorption is related to the chiral effect on the vibration and electronic transitions. In Raman optical activity (ROA), this property is related to the chiral effect on the Raman virtual state and vibration.

Circular dichroism (CD) is the general terminology for this activity. In IR absorption, it is called vibrational circular dichroism (VCD). In the Raman process, it is called ROA [1].

An ultraviolet CD spectrum is easier to obtain than VCD and ROA because of its eminent effect. The ROA experiment is more difficult than VCD because ROA signal is very weak, that is, only 10^{-4} of Raman while that of VCD is 10^{-3}. Currently, commercial CD and VCD instruments are popular, but not the ROA instrument.

The production of right and left circularly polarized light is crucial. Circularly polarized light originates from two orthogonally linearly polarized light beams with phase difference of a quarter wavelength. This can be realized by applying pressure or voltage on photoelastic (ZnSe) or photoelectric (KDP at 2 kV) material so that the components of the passing light beam along different polarizations can be phase lagged of $\pi/2$. $\lambda/4$ plate with such a property can also be used to produce circularly polarized light. Since VCD or ROA signal is very small, eliminating artifacts is the crucial factor in their measurements. This includes the stability control of temperature, to avoid the use of lens and glass as possible, alternatively taking signals by right and left circularly polarized light during the long sampling time, if necessary.

The progress in microelectronics, CCD and controlling software is rapid and is making VCD and ROA routine techniques, though not yet mature, in the past 30 years. We have to recognize their potential roles in molecular spectroscopy, especially ROA.

https://doi.org/10.1515/9783110625097-011

Figures 11.1 and 11.2 show the VCD [2] and ROA [3] spectra. The spectra are defined as follows:

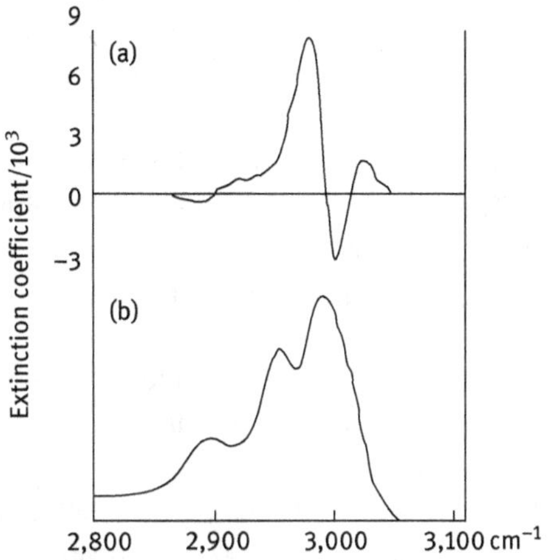

Fig. 11.1: (a) VCD C–H spectrum of L-alanine in D_2O solution and (b) its IR spectrum.

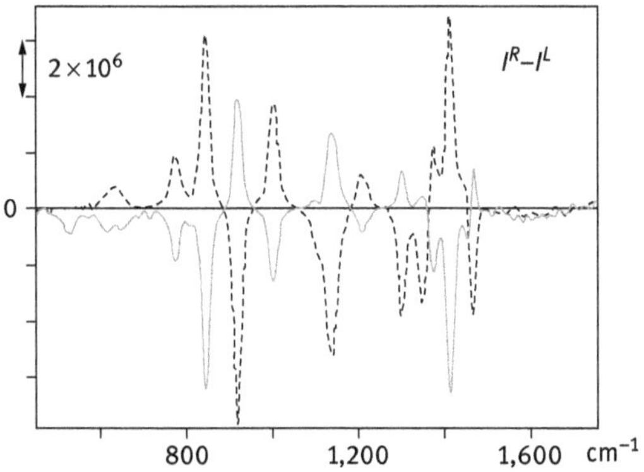

Fig. 11.2: ROA spectrum of L-alanine. Dashed line shows the R-alanine spectrum. The two spectra are of opposite signature, showing no artifacts.

$A^R - A^L$, $I^R - I^L$ with A, I is the extinction coefficient and Raman intensity. L, R refer to the left and right circularly polarized light experiments. VCD and ROA mode intensities and those of IR and Raman are irrelevant. The electronic CD mode spectral peak usually splits into two parts: one is positive and the other is negative. This is because of the term $(\nu^2 - \nu_0^2)^{-1}$, with ν_0 as the mode frequency and ν as the spectral frequency. This is the Cotton effect [4].

11.2 Coupling of electric dipole with magnetic dipole and electric quadrupole

The chiral effect is because of coupling between electric dipole with magnetic dipole and electric quadrupole. This is the second-order process.[1] The IR process is a first-order process. The Raman process is a second-order process. It involves a two-photon process.

The dipolar coupling $\boldsymbol{\mu}\cdot\boldsymbol{\mu}$ is invariant under mirror reflection and center inversion. It does not produce a chiral effect. The coupling of electric dipole with magnetic dipole, $\boldsymbol{\mu}\cdot\boldsymbol{m}$, and that with electric quadrupole, $\boldsymbol{\mu}\cdot\overleftrightarrow{\boldsymbol{Q}}$, are not invariant under mirror reflection and center inversion. ($\overleftrightarrow{Q}$ is the quadrupole. It is a second-rank tensor.)

Suppose $\boldsymbol{\mu}$, $\boldsymbol{\mu}'$, $\boldsymbol{m}$ and $\boldsymbol{m}'$ are the images under mirror reflection and inversion. Figures 11.3(a) and (b) show $\boldsymbol{\mu}\cdot\boldsymbol{m} = -\boldsymbol{\mu}'\cdot\boldsymbol{m}'$ under mirror reflection and inversion. Note that $\boldsymbol{m}$ is a pseudovector since it originates from the electric current.

For $\boldsymbol{\mu}\cdot\overleftrightarrow{\boldsymbol{Q}}$, we have similar transformation. $\overleftrightarrow{\boldsymbol{Q}}$ is a tensor. Its component, say $\overleftrightarrow{\boldsymbol{Q}}_{xy}$, is proportional to xy, which is invariant under inversion. Hence, $\overleftrightarrow{\boldsymbol{Q}}$ is an invariant. Therefore, we have $\boldsymbol{\mu}\cdot\overleftrightarrow{\boldsymbol{Q}} = -\boldsymbol{\mu}'\cdot\overleftrightarrow{\boldsymbol{Q}}'$. For mirror reflection, the situation is more complicated. However, the transformation property is preserved.

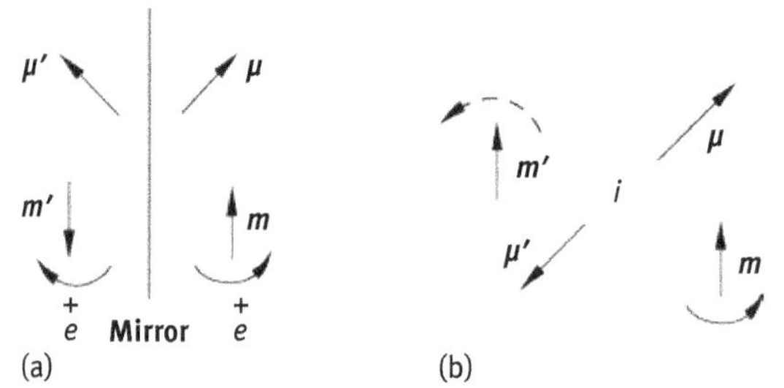

Fig. 11.3: The transformation of electric and magnetic dipoles under mirror reflection and center inversion.

It is then clear that molecular chirality is because of the non-invariance of $\boldsymbol{\mu}\cdot\boldsymbol{m}$ and $\boldsymbol{\mu}\cdot\overleftrightarrow{\boldsymbol{Q}}$ under mirror reflection and center inversion. These couplings are second order, so are VCD and ROA processes. They involve deeper mechanisms of interaction between molecule and light than that of IR and Raman effects. Therefore, VCD or ROA is the unavoidable choice for the study of molecular spectroscopy.

1 The order of the processes involving magnetic dipole and electric quadrupole is higher than that involving electric dipole by $(2\pi/\lambda \cdot a)^2$ with a as the molecular dimension, λ represents the light wavelength. For $a = 10\,\mathring{A}$, $\lambda = 5,000\,\mathring{A}$, then $(2\pi/\lambda \cdot a)^2 \approx 10^{-4}$ (Refer to Exercise.1.3)

11.3 Models for vibrational dichroism

We have known that $\boldsymbol{\mu}\cdot\boldsymbol{m}$ and $\boldsymbol{\mu}\cdot\overleftrightarrow{\boldsymbol{Q}}$ are the origin of vibrational dichroism. These quantities can be calculated based on quantal algorithm. CD, VCD or ROA spectra can be simulated to compare with the experimental results. Though the algorithm and software are rather mature nowadays, they are not absolutely dependable. Furthermore, these software programs are like black boxes in which outputs do not immediately offer clear physical pictures for the systems studied. Therefore, various simplified models were proposed. These models approximately attach $\boldsymbol{\mu}$, $\boldsymbol{m}$, $\overleftrightarrow{\boldsymbol{Q}}$ on atoms, bonds or molecules. These result in the so-called atomic model, bond model and molecular orbital model.

Chirality can originate from the chiral group in a molecule or an achiral molecule in a chiral environment.

11.4 Vibrationally induced charge flow model

The bond vibration will induce an electric dipole, $\boldsymbol{\mu}$. The vibration will also induce charge flow and produce magnetic dipole, $\boldsymbol{m}$. For a chiral molecule, $\boldsymbol{\mu}\cdot\boldsymbol{m}$ is nonzero. (For an achiral molecule, electric and magnetic dipoles can exist locally, but as a whole for the molecule, $\boldsymbol{\mu}\cdot\boldsymbol{m}$ is zero.) It is convenient to understand VCD and ROA by this concept. Sometimes, this concept may help predict the sign of $\boldsymbol{\mu}\cdot\boldsymbol{m}$ and even the spectral signature of VCD and ROA. The absolute molecular configuration may then be determined. This is the charge flow model.The following cases demonstrate this idea [5].

We first determine the orientation of $\boldsymbol{\mu}$ and $\boldsymbol{m}$ as the chiral C–H bond vibrates. As shown in Fig. 11.4, when H atom moves toward the chiral carbon atom (i.e., C–H bond contracts), the induced $\boldsymbol{\mu}$ is shown therein. (Note the definition is with respect to positive charge, so is the electric current.) Meanwhile, if the C–X bond is weaker than C–R_2, the electron flow caused by the contracting C–H bond will flow toward C–X and X–H. The induced magnetic dipole is shown therein. Since the angle between $\boldsymbol{\mu}$ and $\boldsymbol{m}$ is less than 90°, $\boldsymbol{\mu}\cdot\boldsymbol{m} > 0$. When the C–H bond expands, the orientation of $\boldsymbol{\mu}$ and $\boldsymbol{m}$ are opposite but still $\boldsymbol{\mu}\cdot\boldsymbol{m} > 0$. The electron flow is more eminent when there is a complete cycle path by the hydrogen bonding formation of R_2. . .H.

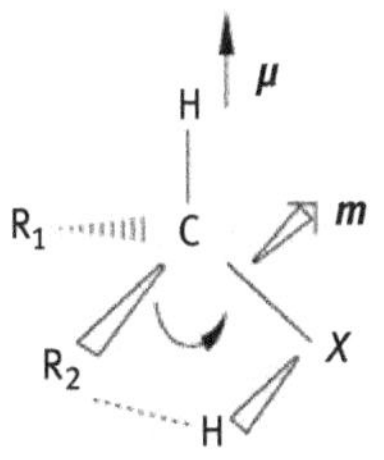

Fig. 11.4: Induced μ and $\boldsymbol{m}$ as H moves toward C atom.

Based on this concept, we can infer the orientation of $\boldsymbol{\mu}$ and m for the configurations shown in Fig. 11.5 for (a) $\boldsymbol{\mu}\cdot\boldsymbol{m} > 0$ and (b) $\boldsymbol{\mu}\cdot\boldsymbol{m} < 0$. The experiment shows that $\boldsymbol{\mu}\cdot\boldsymbol{m} > 0$. Hence, (a) is the preferred configuration.

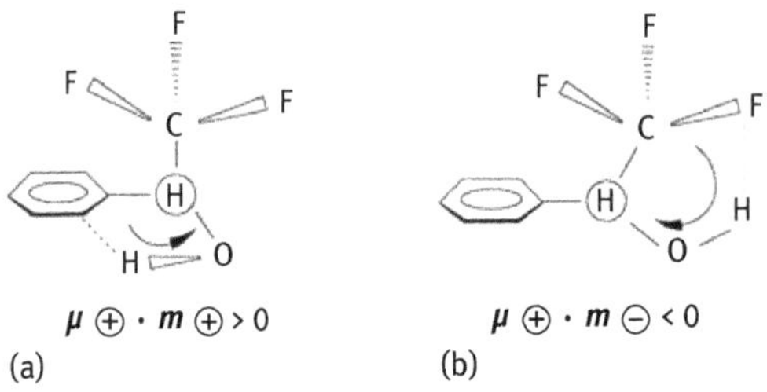

Fig. 11.5: C–H vibrationally induced μ and $\boldsymbol{m}$. ⊕ and ⊖ signs show out of and into the plane.

The molecule shown in Fig. 11.6 has two configurations in equilibrium. For (a), $\boldsymbol{\mu}\cdot\boldsymbol{m} > 0$, while for (b), $\boldsymbol{\mu}\cdot\boldsymbol{m} < 0$. The experiment shows $\boldsymbol{\mu}\cdot\boldsymbol{m} > 0$. Hence, (a) is the more probable configuration.

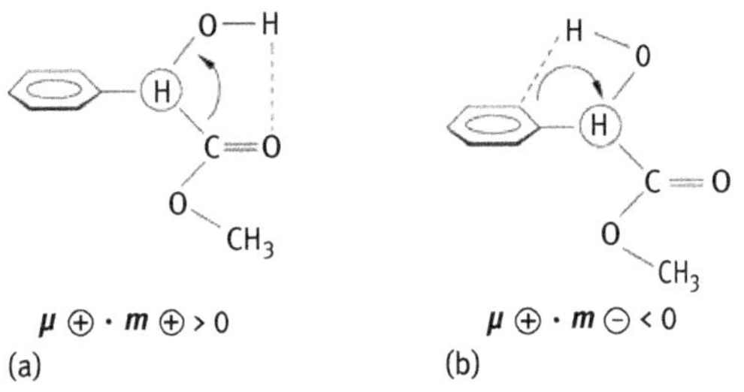

Fig. 11.6: Different structure corresponds to different $\mu\cdot\boldsymbol{m}$ signs.

Figure 11.7 shows two configurations. As the O–H bond stretches out, the hydrogen bond between –C–O. . .H–O– contracts so that the electrons on –C–O flow toward the hydrogen bond, resulting in the induced $\boldsymbol{\mu}$ and $\boldsymbol{m}$ as shown therein.

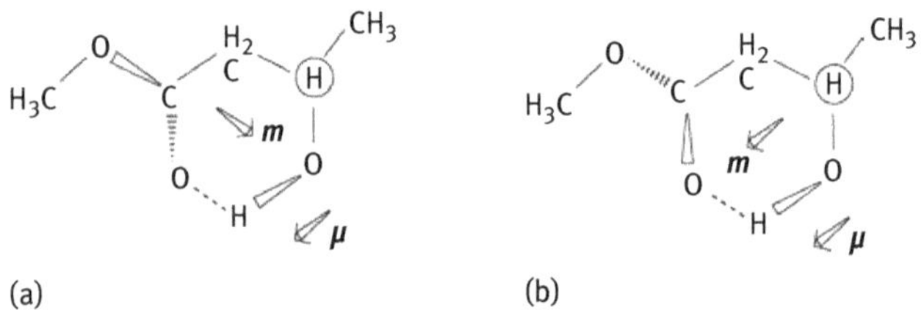

Fig. 11.7: Different structures correspond to different $\mu\cdot\boldsymbol{m}$ signs.

For (a) $\boldsymbol{\mu}\cdot\boldsymbol{m} < 0$. For (b), $\boldsymbol{\mu}\cdot\boldsymbol{m} > 0$. If the sign of $\boldsymbol{\mu}\cdot\boldsymbol{m}$ can be inferred from VCD/ROA spectrum, the absolute configuration can be determined.

11.5 Conclusion

VCD and ROA are effective tools in determining the absolute configurations for biological molecules. A biosystem is usually present in an aqueous environment. VCD is opaque for spectral region below 500 cm^{-1} (mainly related to the molecular backbone vibration). Hence, ROA is superior in biosystem and shows broad and potential perspectives.

Chirality is a fundamental concept in physics. People in two isolated worlds can make sure that their definition of "right" and "left" is consistent by measuring VCD spectra of Fig. 11.5(a) and having positive $\boldsymbol{\mu}\cdot\boldsymbol{m}$, so that their A^R and A^L are in agreement.

References

[1] Barron LD. Molecular Light Scattering and Optical Activity. United Kingdom: Cambridge University Press, 1982.
[2] Lal BB, Diem M, Polavarapu PL, Oboodi MR, Freeman TB, Nafie LA. J. Am. Chem. Soc. 1982, 104: 3336.
[3] Fang Y, Wu G, Wang P. Chem. Phys. 2012, 393: 140.
[4] Charney E. The Molecular Basis of Raman Optical Activity. New York:Wiley, 1979.
[5] Freedman TB, Balukjian GA, Nafie LA. J. Am. Chem. Soc. 1985, 107: 6213.

12 Differential bond polarizability and Raman optical activity

12.1 Bond polarizability in Raman optical activity

As a molecule cannot be superimposed on its mirror image, it attains chirality/handedness. A chiral molecule will possess differential Raman intensities under the right and left circularly polarized laser excitations. The intensity difference is very small, only 10^{-4} of its Raman intensity. This is called Raman optical activity (ROA). Similar phenomenon occurs in the electronic spectroscopy, which is called circular dichroism (CD) and in IR it is the so-called vibrational circular dichroism (VCD).

In Chapter 10, we discussed the study of the Raman virtual state. Our method aims at obtaining bond polarizabilities from the Raman mode intensities. In fact, this method can be easily extended to ROA. We consider the differential Raman intensities, $I_j^R - I_j^L = \Delta I_j$ (where R and L stand, respectively, for the right and left circularly polarized excitations and ΔI_j, which is the difference of the Raman intensities by these two excitations, is called the ROA mode intensity) and try to elucidate the corresponding differentiation of the bond polarizability $\partial\Delta\alpha/\partial S_k$ (where S_k is the internal coordinate). This differentiation of the bond polarizability provides considerable ROA information. We call $\partial\Delta\alpha/\partial S_k$ the differential bond polarizability.

In Chapter 10, we discussed the method to obtain the bond polarizabilities from a set of Raman mode intensities. The equation set is given as follows:

$$\begin{bmatrix} \partial\alpha/\partial S_1 \\ \partial\alpha/\partial S_2 \\ \\ \partial\alpha/\partial S_k \\ \vdots \\ \partial\alpha/\partial S_t \end{bmatrix} = \begin{bmatrix} \\ a_{jk} \\ \\ \end{bmatrix}^{-1} \begin{bmatrix} P_1\sqrt{I_1} \\ P_2\sqrt{I_2} \\ \\ P_j\sqrt{I_j} \\ \vdots \\ P_t\sqrt{I_t} \end{bmatrix}$$

with $a_{jk} = \left(\nu_0 - \nu_j\right)^2/\sqrt{\nu_j}(L^T)_{jk}$, ν_0 the laser wave number, ν_j the Raman shift, $\partial S_k/\partial Q_j = L_{kj}$ and Q_j the normal coordinate. P_j can be positive or negative. (Note that only relative intensities and polarizabilities are considered.)

For the ROA experiment, we have $I_j^R + I_j^L = I_j$ and $I_j^R - I_j^L = \Delta I_j$,

$$\text{or } I_j^R = (I_j + \Delta I_j)/2 \text{ and } I_j^L = (I_j - \Delta I_j)/2$$

https://doi.org/10.1515/9783110625097-012

Then we have the following:

$$\begin{bmatrix} \partial\Delta\alpha/\partial S_1 \\ \partial\Delta\alpha/\partial S_2 \\ \vdots \\ \partial\Delta\alpha/\partial S_t \end{bmatrix} = \begin{bmatrix} & & \\ & a_{jk} & \\ & & \end{bmatrix}^{-1} \begin{bmatrix} P_1(\sqrt{I_1^R} - \sqrt{I_1^L}) \\ P_2(\sqrt{I_2^R} - \sqrt{I_2^L}) \\ \vdots \\ P_t(\sqrt{I_t^R} - \sqrt{I_t^L}) \end{bmatrix}$$

Here, $\Delta\alpha$ corresponds to $\alpha^R - \alpha^L$.
Consider that

$$\sqrt{2I_j^R} = \sqrt{I_j + \Delta I_j} = \sqrt{I_j}[\sqrt{(1 + \Delta I_j/I_j)}] \approx \sqrt{I_j}[1 + \Delta I_j/2I_j],$$

$$\sqrt{2I_j^L} = \sqrt{I_j - \Delta I_j} = \sqrt{I_j}[\sqrt{(1 - \Delta I_j/I_j)}] \approx \sqrt{I_j}[1 - \Delta I_j/2I_j],$$

with an error $< [\Delta I_j/I_j]^2/8$

then

$$\sqrt{I_j^R} - \sqrt{I_j^L} \approx \Delta I_j/\sqrt{I_j}$$

and

$$\begin{bmatrix} \partial\Delta\alpha/\partial S_1 \\ \partial\Delta\alpha/\partial S_2 \\ \vdots \\ \partial\Delta\alpha/\partial S_t \end{bmatrix} = \begin{bmatrix} & & \\ & a_{jk} & \\ & & \end{bmatrix}^{-1} \begin{bmatrix} P_1(\Delta I_1/\sqrt{I_1}) \\ P_2(\Delta I_2/\sqrt{I_2}) \\ \vdots \\ P_t(\Delta I_t/\sqrt{I_t}) \end{bmatrix}$$

Although I_j is much larger than ΔI_j by an order of 10^3–10^4, the relative magnitudes of I_js and ΔI_js can be treated independently as not too small numbers. (For instance, they can be scaled to from 1 to 100 in most cases.) By this way, $\Delta I_j/\sqrt{I_j}$ can be treated as not too small numbers. We also note that since the intensity difference between the right and left circularly polarized excitations is so small, only of the order of 10^{-3}–10^{-4} of the Raman intensities, their phases are the same.

Here, we note that those smaller Raman mode intensities and larger ROA intensities will contribute more to the differential bond polarizabilities. These modes will provide more ROA information.

Once the bond polarizabilities were obtained from I_j's, then together with the elucidated P_j's and $\Delta I_j'$s that are obtained from the ROA experiment, relative $\partial\Delta\alpha/\partial S_k$ can be obtained.

$\partial\Delta\alpha/\partial S_k$ provides the ROA information such as the vibrationally induced magnetic and electric quadrupole effects. All these cannot be obtained from Raman and IR spectra.

The chiral center plays the key role in ROA. For those parts that are farther away from the chiral center, like the peripheral C–H (in high wave number region), their corresponding ROA spectra will be very small or even vanishing, that is, their $\Delta I_j = 0$.

In the ROA experiment, to obtain accurate I_j and ΔI_j is important. The response correction of grating mirror and CCD has to be considered. The temperature stability is also important since it may affect the polarization quality of the laser excitation. Less lens or glass is crucial in avoiding artifacts.

We note that Raman provides information regarding bond force field and virtual state; however, it does not provide more steric information as ROA does.

12.2 Bond polarizability and differential bond polarizability in (+)-(R)-methyloxirane

Figure 12.1 shows the structure of methyloxirane and its atomic numberings. It contains no nontrivial symmetry.

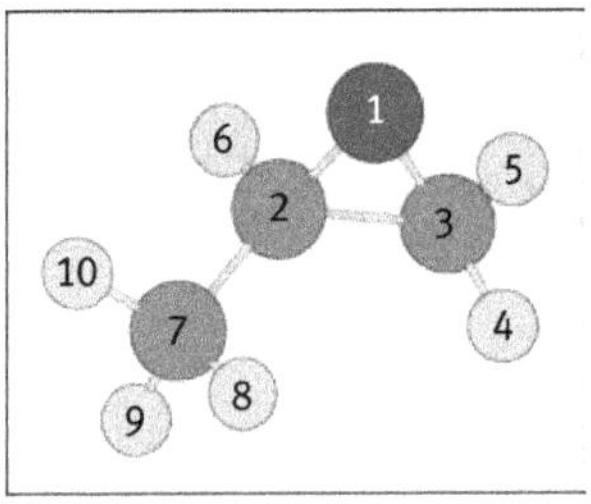

Fig. 12.1: The structure of (+)-(R)-methyloxirane and its atomic numberings. 1 is oxygen atom and 2, 3 and 7 are carbon atoms. The rest are hydrogen atoms.

From its Raman mode intensities (under 532 nm excitation), the bond polarizabilities can be obtained as shown in Fig. 12.2. (In determination of the phase, the solution is not unique. However, they offer quite consistent bond polarizabilities. Here, only the representative one is shown.)

From Fig. 12.2, we note that at the initial stage of excitation, the bond polarizabilities of C2–H6, C2–C3, C2–O1 and C3–O1 are noticeably larger than that of C2–C7, while for the ground state, C2–C7 possesses the largest electronic density. This implies that in the Raman process, there is the coupling of dipole on C2–H6 and the magnetic moment that originates from the charge flow on the C–C–O skeleton. This plays the role for ROA. Figure 12.3 demonstrates this process.

With this picture, we expect that the modes with more components of the stretching/bending motion of the triangular oxirane skeleton will possess more significant ROA activity. Indeed, the ROA spectrum shown in Fig. 12.4 shows that the modes about 898 cm^{-1}, 832 cm^{-1} and 750 cm^{-1}, which are mostly of the stretching/bending character of the triangular oxirane as shown by their PED (potential energy distribution), possess the most significant ROA activity. Also note that the mode

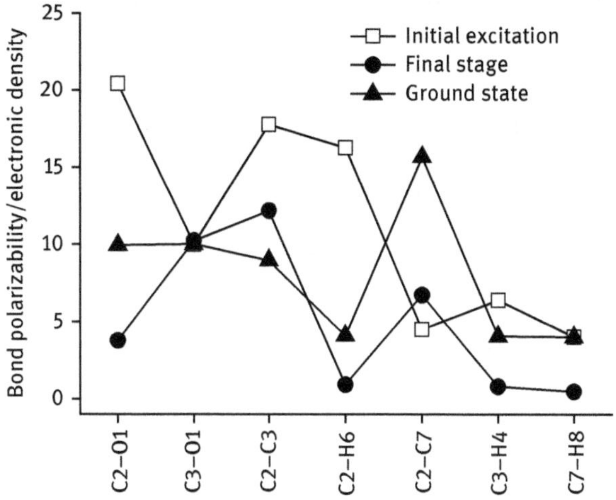

Fig. 12.2: The relative bond (stretch) polarizabilities of (+)-(*R*)-methyloxirane at the initial excitation moment (□) and the final stage of Raman relaxation (●) by 532 nm excitation. (▲) is for the calculated bond electronic densities of the ground state. For convenience, the values of C3–O1 are normalized to 10. Note that there is no correlation between the values of the bond polarizability and the calculated bond electronic density.

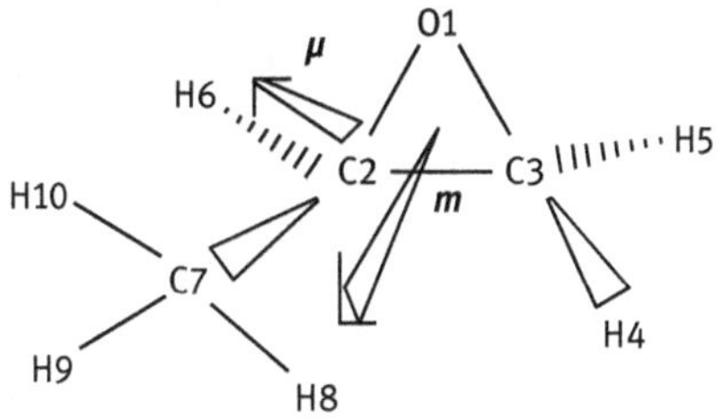

Fig. 12.3: The sketch of a significant vibrating electronic dipole μ induced by the excited charges on the C2–H6 bond and an eminent induced magnetic moment m on the triangular oxirane skeleton near the asymmetric center as the skeleton vibrates in the Raman excitation. Their coupling results in the significant ROA for the skeletal modes. Note that there are bare excited charges on the C2–C7 bond in the Raman excitation so that its motion will not contribute to ROA, though it possesses the most significant electronic density in the ground state.

about 953 cm^{-1} does not show significant ROA activity (though of Raman significance) since it is more of the stretching motion along the C2–C7 bond and the bending motion associated with C2–C7–H8 (–H9, –H10). Evidently, the Raman chirality of this compound does not come from the dipole on the C2–C7 bond, though it possesses the largest bond electronic density in the ground state.

In summary, to elucidate the bond polarizabilities from the Raman mode intensities is helpful for the understanding of ROA. The point is that, during the

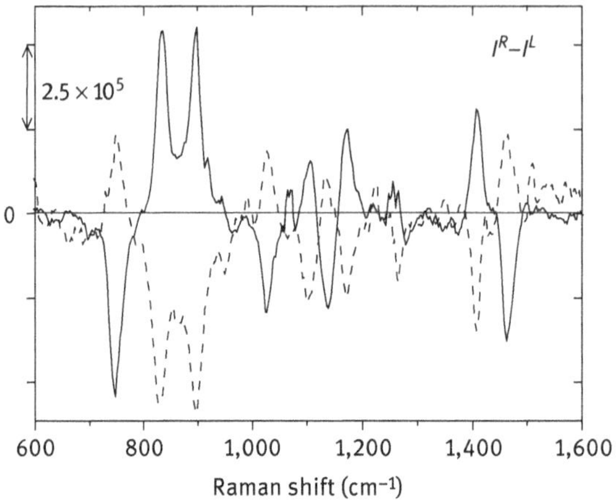

Fig. 12.4: The ROA spectra of (+)-(R)-methyloxirane by 532.5 nm excitation. Those of (–)-(S)-methyloxirane are also shown by dashed lines. They are opposite, showing that there are no artifacts.

Raman process, the excited charges are aggregated toward the asymmetric center C–H bond, which is of bare charges in the ground state. The coupling of this induced large dipole with the eminently induced magnetic moment around the nearby structure thus leads to the Raman chirality. This is common in many chiral molecules.

Once the phases are determined, with Raman and ROA mode intensities, the differential bond polarizabilities can be obtained as shown in Fig. 12.5. We note that different phase sets result in quite consistent results.

It is impressive to note that the differential bond polarizabilities of the C2–C3 and C2–C7 stretching coordinates, which are on the two opposite sides along the C2–H6 bond, are of opposite signs. Moreover, we note that those of the bending coordinates associated with C7 (such as O1–C2–C7,C2–C3–C7 (S14) and C7–H_3 (S23)) and C3 (O1–C3–H_2 and C2–C3–H_2 (S15)) are also of opposite signs. This means that the responses of the portions, on the two opposite sides along the C2–H6 bond, to the right and left circularly polarized excitations are in opposite ways. This is probably due to the opposite orientations of the vibrationally induced charge currents along the oxirane ring and the C2–C7–H_3 paths in the Raman process.

Finally, from Fig. 12.2, we note that the bond polarizabilities after relaxation is consistent with the electronic bond densities in the ground state calculated by all the occupied MOs. This means that all the electrons in the molecule are involved in the Raman process. This is meaningful and significant if one recognizes that *conceptual* bond density can now be related to the bond polarizability derived from the *experiment* (refer to Section 10.3)

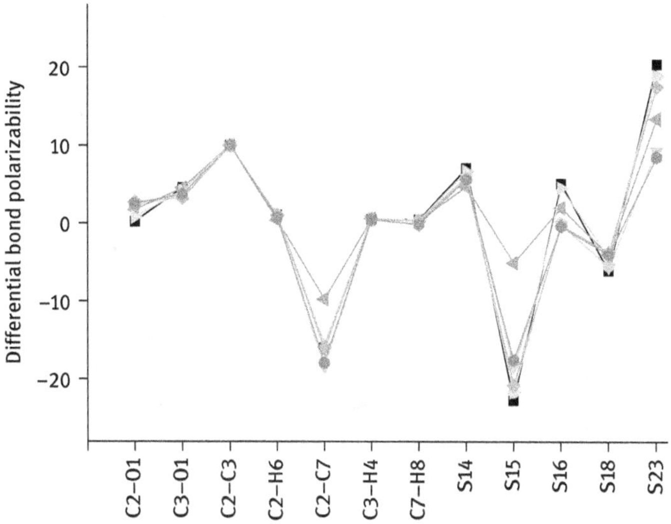

Fig. 12.5: The relative differential bond polarizabilities of (+)-(R)-methyloxirane under 532.5 nm excitation by nine-phase set solutions. For convenience, the values of C2–C3 are normalized to 10. S14 involves the O1–C2–C7 and C2–C3–C7 bendings, S15 involves the O1–C3–H_2 and C2–C3–H_2 bendings, S23 involves the C7–H_3 bending and S16 and S18 involve the C3–H_2 wagging (bending).

12.3 Intramolecular enantiomerism

By differential bond polarizability study, we found the so-called intramolecular enantiomerism. We know that chiral right and left enantiomers, which are images to each other through a mirror reflection, possess identical ROA spectra, but with opposite signs. Hence, their corresponding differential bond polarizabilities are the same but with opposite signs. However, we found that chiral molecules with ring structure possess *approximate* mirror symmetry, which of course is not strict; otherwise, the molecules will be achiral. Although the mirror-related bonds possess variant differential bond polarizabilities, they are of opposite signs, just like that between the chiral right and left enantiomers. The mirror symmetry breaks due to the nonidentical structures of the two mirror-related parts of the molecule. The symmetry breaking is not small perturbation; however, the *chiral imaging* is still preserved as far as the sign of differential bond polarizability is concerned. It is called as the intramolecular enantiomerism.

Figures 12.6 and 12.7 show this phenomenon in R-limonene and 2,2-dimethyl-1,3-dioxolane-4-methanol. The mirror symmetry by the signs of the differential bond polarizabilities is obvious.

The magnitude difference between the differential bond polarizabilities of the pair coordinates related by the intramolecular mirror reflection is an indication of the degree of ROA. It shows the core of ROA phenomenon. The larger the difference, the more eminent is ROA and so is the larger the mirror symmetry breaking.

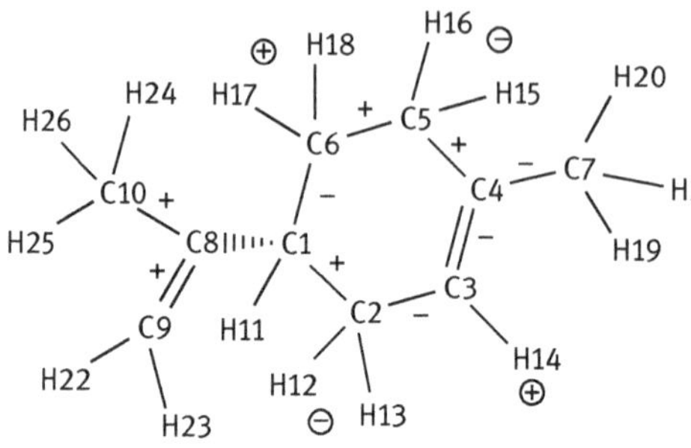

Fig. 12.6: The signs of the differential bond polarizabilities for R-limonene. (+, –) for C–C stretching coordinates and (⊕), (⊖) for the bending coordinates (C6–H_2 bending, C2–H_2 bending, C5–H_2 bending, C3–H bending). The mirrors are vertical to the molecular plane, connecting C1 and C4, and the midpoints of C2–C3 and C5–C6 bonds.

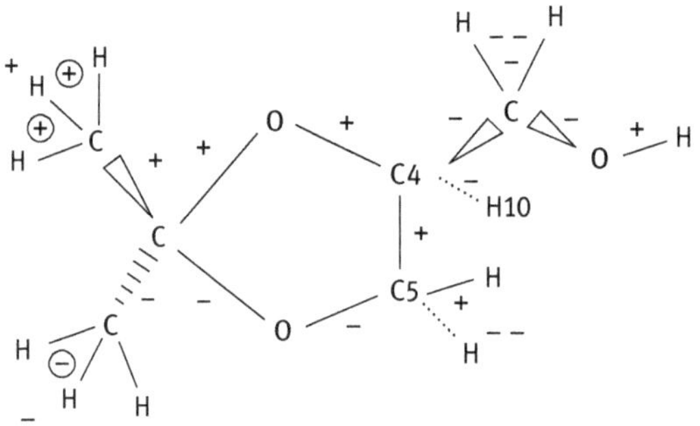

Fig. 12.7: The signs of the differential bond polarizabilities of 2,2-dimethyl-1,3-dioxolane-4-methanol. (+), (–) for C–C, C–O, O–H bond stretchings and C–H_3, C–H_2 symmetric stretching coordinates.(– –)for the C–H_2 antisymmetric stretching coordinate. (⊕), (⊖) for the C–H_3 bending coordinates. The mirror is vertical to the molecular plane, connecting the chiral atom and the midpoint of C4–C5 bond. Two C–H_3 moieties are on the two sides of the ring plane. They show definite enantiomerism.

Figures 12.8(a) and (b) shows this difference for S-2,2-dimethyl-1,3-dioxolane-4-methanol and R-limonene. We note that at the chiral center the symmetry breaking is the largest and as the pair bond coordinates are farther away from the asymmetric atom, this magnitude difference becomes smaller. The role of the chiral center in ROA is confirmed. We stress that this is a vivid spectroscopic demonstration of the asymmetry of the chiral center.

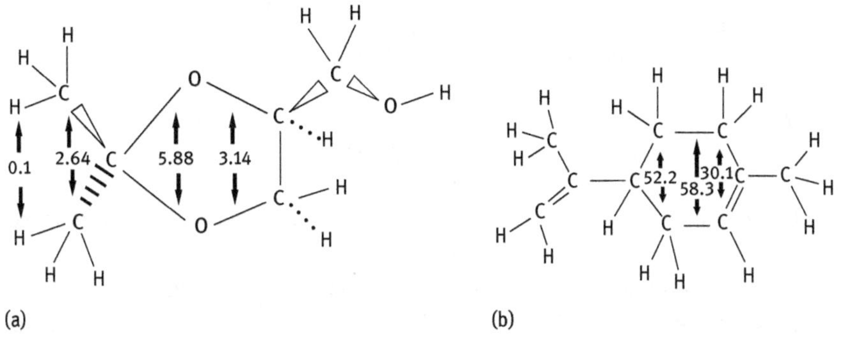

Fig. 12.8: The magnitude differences between the differential bond polarizabilities of the pair coordinates related by the intramolecular mirror reflection. (a) 2,2-dimethyl-1,3-dioxolane-4- methanol, for the C–O, C–C and C–H_3 symmetric stretching coordinates. (b) (R)-limonene for the C–C stretching coordinates.

12.4 Equivalence between Raman, ROA intensities and bond polarizability, differential bond polarizability

We know the following relation:

$$\begin{bmatrix} \partial\alpha/\partial S_1 \\ \partial\alpha/\partial S_2 \\ \vdots \\ \partial\alpha/\partial S_t \end{bmatrix} = \left[a_{jk}\right]^{-1} \begin{bmatrix} P_1\sqrt{I_1} \\ P_2\sqrt{I_2} \\ \vdots \\ P_t\sqrt{I_t} \end{bmatrix}$$

This transformation is linear. Bond polarizability and Raman mode intensity are, therefore, equivalent. From the experiment, we have the Raman intensities. Bond polarizabilities are the molecular parameters. However, there is a difference. Bond polarizability is extracted from the Raman intensity without the contents that are irrelevant to the Raman scattering, such as L_{kj}, which is related to the molecular configuration, atomic mass and force field. We thus understand that bond polarizability bears the core information of the Raman process.

Similarly, there is linear relation between differential bond polarizabilities and ROA mode intensities (scaled by [Raman intensity]$^{1/2}$):

$$\begin{bmatrix} \partial\Delta\alpha/\partial S_1 \\ \partial\Delta\alpha/\partial S_2 \\ \vdots \\ \partial\Delta\alpha/\partial S_t \end{bmatrix} = \left[a_{jk}\right]^{-1} \begin{bmatrix} P_1(\Delta I_1/\sqrt{I_1}) \\ P_2(\Delta I_2/\sqrt{I_2}) \\ \vdots \\ P_t(\Delta I_t/\sqrt{I_t}) \end{bmatrix}$$

They are the two facets of a physical quantity. Differential bond polarizability is extracted from the ROA intensity only those contents that are relevant to ROA. In this sense, differential bond polarizability bears additional Raman information.

Finally, we have to recognize that *all* the intensities in the scattered light regardless of their different mechanisms, if there are, are included in the intensity we employ for analysis. In this sense, the terminology *polarizability* used is just for convenience. Any overinterpretation of *polarizability* should be avoided.

12.5 A classical theory for ROA and charge distribution in Raman virtual state

The main physical mechanism of ROA is the coupling of vibrationally induced electric and magnetic dipoles. This physical picture of ROA mechanism is analyzed in a classical way.

The jth atom in a molecule with net charge e_j, under a normal mode vibration will oscillate around its equilibrium position periodically. Its displacement $\Delta\boldsymbol{r}_j$ and the induced dipole $\boldsymbol{\mu}_j$ (i.e., $e_j\Delta\boldsymbol{r}_j$) are coincident in space. Meanwhile, the other ith atom, with net charge e_i, will produce a magnetic moment $\boldsymbol{m}_{ij}$ on the jth atom, with a value of

$$\frac{1}{2}(e_i\boldsymbol{v}_i \times \boldsymbol{r}_{ij})$$

Here, $\boldsymbol{r}_{ij}$ is the position vector from the ith to the jth atoms. $\boldsymbol{v}_i$ is the velocity of the ith atom.

The quantal expression for the interaction between the vibrationally induced electric and magnetic dipoles is given as follows [1]:

$$Im(\langle i|\boldsymbol{\mu}|f\rangle\langle f|\boldsymbol{m}|i\rangle)$$

Here, $\langle i|$ and $|f\rangle$ are the initial and final states. Im shows the imaginary part of the electromagnetic interaction, for which $\boldsymbol{\mu}$ and $\mathbf{m}$ are out of phase by $\pi/2$. Hence, its classical expression is as follows:

$$\sum \mu \bullet m$$

Specifically, for a normal mode, the term that determines its Δ_{cl} sign is as follows:

$$\sum_j \sum_i \boldsymbol{\mu}_{j\max} \bullet \boldsymbol{m}_{ijequ}$$

Here, $\boldsymbol{\mu}_{j\max}$ is the dipole when $\Delta \boldsymbol{r}_j$ reaches its maximum (denoted as $\Delta \boldsymbol{R}_j$). $\boldsymbol{m}_{ijequ}$ is the magnetic moment induced by the motion of the charge on the ith atom at the jth atom when $\Delta \boldsymbol{r}_i = 0$ with maximal $\boldsymbol{v}_i$. (Note the phase difference of $\pi/2$ between $\boldsymbol{\mu}_{j\max}$ and $\boldsymbol{m}_{ijequ}$.) When $\Delta \boldsymbol{r}_i = 0$, $\boldsymbol{v}_i$ is proportional to $\Delta \boldsymbol{R}_i$. (As $\Delta \boldsymbol{r}_i = 0$, $\boldsymbol{v}_i$ is proportional to $\Delta \boldsymbol{R}_i/\tau$. τ is the period of the normal mode.) Hence, Δ_{cl} is determined by the sign of the following expression:

$$\begin{aligned} &\sum_{ji} (e_j \Delta \boldsymbol{R}_j) \bullet (e_i \Delta \boldsymbol{R}_i \times \boldsymbol{r}_{ij}) \\ &= \sum_{ji} (e_i e_j \boldsymbol{r}_{ij}) \bullet (\Delta \boldsymbol{R}_j \times \Delta \boldsymbol{R}_i) \end{aligned}$$

Since $|\Delta \boldsymbol{R}_i|$ ($|\Delta \boldsymbol{R}_j|$) is much smaller than $|\mathbf{r}_{ij}|$, $\mathbf{r}_{ij}$ can take the value at the molecular equilibrium configuration. In principle, $\mathbf{r}_{ij}$ can be obtained from the molecular configuration or by the quantum chemical calculation. $\Delta \boldsymbol{R}_i$ ($\Delta \boldsymbol{R}_j$) can be obtained from the normal mode analysis, if force constants are known. As to the chirality of the aforementioned formula, we note that $\mathbf{r}_{ij} \bullet (\Delta \boldsymbol{R}_j \times \Delta \boldsymbol{R}_i)$ changes its sign as the coordinate system is shifted between the right-hand one and the left-hand one as anticipated.

The charge e_j for VCD is the Mullikan charge e_j^0, while for ROA, it is the one we derived in Section 10.5:

$$e_j = e_j^0 + t \cdot z_j \cdot \sum_l \partial\alpha/\partial R_l/N - t \sum_{kj} \partial\alpha/\partial R_{kj,j}/2$$

We thus put VCD and ROA mechanisms on equal footing. In the following, we first consider the ROA case.

The molecular system we will try on is (+)-(R)-methyloxirane whose structure together with its atomic numberings is shown in Fig. 12.1. From its Raman (relative) intensities, the (relative) bond polarizabilities were determined under 532 nm excitation. The relative bond polarizabilities with that of C2O1 normalized to 100 are listed in Tab. 12.1. The Mulliken charges (e_j^0) were calculated by the DFT quantum method based on 6-31G* basis. For this molecule, $\sum_l \partial\alpha/\partial R_l = 459.64$ as the bond polarizability of C2O1 is normalized to 100. $N = 32$ ($N = 6 \cdot 3 + 8 \cdot 1 + 1 \cdot 6 = 32$) and t has to be less than 0.0696. (Note that $t \le N/\sum_l \partial\alpha/\partial R_l$; see Section 10.5). Other parameters were also calculated and tabulated in Tab. 12.1. The calculation of $z_j \sum_l \partial\alpha/\partial R_l/N - \sum_{kj} \partial\alpha/\partial R_{kj,j}/2$ (see Section 10.5) shows that in the Raman excitation, C2, C3, H4, H5 and H6, especially, C2 and H6, acquire electrons, while the rest of the atoms, especially, C7 and O1, lose electrons.

Then the Δ_{cl} sign of a normal mode can be readily determined by the sign of $\sum_{ji} e_i e_j \boldsymbol{r}_{ij} \cdot (\Delta \boldsymbol{R}_j \times \Delta \boldsymbol{R}_i)$, which can be calculated as soon as the normal mode analysis for this molecule is done, with a presumed t factor. In the calculation for Δ_{cl}, we found that if the scaling factor t is in the range of (0.0133, 0.0160); in calculation,

Tab. 12.1: For (+)-(R)-methyloxirane, the relative bond polarizabilities, the polarizability attributed to each atom ($\sum_{kj} \partial\alpha/\partial R_{kj,j}/2$), the Mulliken charge ($e_j^0$), the electron number ($z_j = Z_j - e_j^0$ with Z_j the atomic number) on each atom based on the Mulliken charge, the redistribution of the charges as evidenced by the bond polarizabilities in the virtual state ($z_j \sum_l \partial\alpha/\partial R_l/N - \sum_{kj} \partial\alpha/\partial R_{kj,j}/2$, where $\sum_l \partial\alpha/\partial R_l = 459.64$ is the sum of all the bond polarizabilities) and the effective charge ($e_j = e_j^0 + t \bullet z_j \bullet \sum_l \partial\alpha/\partial R_l/N - t\sum_{kj} \partial\alpha/\partial R_{kj,j}/2$) range as the scaling factor t varies from 0.0133 to 0.0160.

Bond	**Relative bond polarizability**	**Atom**	$\sum_{k_j} \partial\alpha/\partial R_{k_j,j}/2$	**Mulliken charge** (e_j^0)	z_j	$z_j \sum_l \partial\alpha/\partial R_l/N - \sum_{k_j} \partial\alpha/\partial R_{k_j,j}/2$	$e_j = e_j^0 + t \cdot z_j \cdot \sum_l \partial\alpha/\partial R_l/N - t \sum_{k_j} \partial\alpha/\partial R_{k_j,j}/2$ **with** t **=0.0133 and 0.0160**	
C2O1	100.00	O1	74.49	−0.84	8.84	52.49	−0.15	−0.01
C3O1	48.98	C2	144.46	0.72	5.28	−68.62	−0.18	−0.37
C2C3	87.11	C3	99.47	−0.14	6.14	−11.28	−0.30	−0.33
C2H6	79.80	H4	15.71	0.15	0.85	−3.50	0.11	0.10
C2C7	22.00	H5	15.71	0.16	0.84	−3.65	0.12	0.11
C3H4	31.43	H6	39.90	0.16	0.84	−27.84	−0.20	−0.28
C3H5	31.43	C7	40.45	−0.71	6.71	55.93	0.03	0.18
C7H8	19.63	H8	9.82	0.16	0.84	2.25	0.19	0.20
C7H9	19.63	H9	9.82	0.16	0.84	2.25	0.20	0.20
C7H10	19.63	H10	9.82	0.16	0.84	2.25	0.20	0.21

the digital step is 0.0001, then the experimental ROA signs can be well reproduced by the calculated Δ_{cl}. This is indeed remarkable as shown in Tab. 12.2! Tab. 12.2 also lists the calculated Δ_{cl} signs by the Mulliken charges without those originating from the bond polarizabilities. Then, the discrepancy with the experimental observation is obvious. This indicates that, in the Raman excitation, the charge redistribution is not negligible. The charge redistribution is evidenced by the bond polarizabilities. These charge modifications on the atoms in the Raman excited state are significant. They are of the same order as the Mulliken charges as listed in Tab. 12.1 by the comparison of e_j^0 and

$$t \cdot (z_j \sum_l \partial\alpha/\partial R_l/N - \sum_{k_j} \partial\alpha/\partial R_{k_j,j}/2)$$

Tab. 12.2: For (+)-(R)-methyloxirane, the comparison of ROA spectral signs by the experimental observation, the calculations based on Mulliken charges and the effective charges developed from bond polarizabilities. *N* shows inconsistency to the experimental observation.

Experimental wavenumber	Experimental ROA spectral sign	Calculated Δ_{cl} sign by Mulliken charges	Calculated Δ_{cl} sign by the effective charges as t varies from 0.0133 to 0.0160
1,501.0		−	+
1,477.0	−	−	−
1,458.0		+	−
1,410.0	+	+	+
1,372.0		+	+
1,268.0		−	+
1,170.0	+	*N* −	+
1,147.0		+	−
1,134.0	−	*N* +	−
1,106.0	+	+	+
1,026.0	−	*N* +	−
953.0		−	+
898.0	+	*N* −	+
832.0	+	+	+
750.0	−	*N* +	−

The final determination of the scaling factor t is of significance. It relates the relative bond polarizabilities to the electric charge. Furthermore, we note that, corresponding to $t = 0.0133$ and 0.0160, there are 6.11 and 7.35 ($t \cdot \sum_l \partial\alpha/\partial R_l = t \cdot 459.64$) electrons being excited or involved in the vibronic coupling that leads to the Raman effect. In the methyoxirane molecule, there are 32 electrons; hence, the excitation percentage or probability is between 19% and 23%. That means around 20% of electrons are excited or perturbed in the Raman process under 532 nm excitation.

For VCD of (+)-(R)-methyloxirane, the spectral signs of the VCD peaks by the experiment [2], the quantum chemical calculation by VCT-CO/6-31G* (0,3) [2] and our method based on the Mulliken charges are tabulated in Tab. 12.3. In fact, the disadvantage of our method is not major when compared to the quantum chemical VCT-CO/6-31G* (0,3) calculation; if one notes that by our algorithm based on the Mulliken charges, of the 12 modes, only the two modes at 1,147 and 898 cm^{-1} are of significant inconsistency, and the other three discrepancies occur in the modes at 1,501, 1,170 and 1,106 cm^{-1}, which are of rather small VCD intensities. However, by the quantal method, there are two cases of discrepancy: one is at 1,170 cm^{-1} of very small intensity and the other at 1,134 cm^{-1} of larger intensity. The accuracy of the prediction of VCD spectral signs by our algorithm is noticeable. Indeed, for the VCD process, unlike that of ROA in which the charge excitation in the Raman virtual state is crucial, the Mulliken charges in the ground state play the major role.

In summary, the elucidation of the scaling parameter for relating bond polarizability to electric charge is significant. This is realized through fitting to the ROA signatures by our classical model. This is possible as ROA provides an additional condition to what Raman effect offers. The finding is of significance that around 20% of electrons

Tab. 12.3: The comparison of the VCD spectral signs of (+)-(R)-methyloxirane by the experimental [2], VCT-CO/6-31G*(0,3) calculation [2] and our method based on the Mulliken charges. *N* shows inconsistency to the experimental observation.

Experimental wavenumber	Experimental VCD relative intensity and its sign	Sign by VCT-CO/6-31G* (0,3)	Sign by our method
1,501.0	11	+	*N* –
1,477.0			–
1,458.0			+
1,410.0	51	+	+
1,372.0	18	+	+
1,268.0	−30	–	–
1,170.0	6	*N* –	*N* –
1,147.0	− 36	–	*N* +
1,134.0	53	*N* –	+
1,106.0	−14	–	*N* +
1,026.0	15	+	+
953.0	−100	–	–
898.0	76	+	*N* –
832.0			+
750.0	35	+	+

are excited or perturbed in the Raman process under 532 nm excitation by this simple/classical treatment of ROA. This offers us a concrete impression about the Raman process.

References

[1] Barron LD, Molecular Light Scattering and Optical Activity. Cambridge :UniversityPress, 1982.
[2] Rauk A, Eggimann T, Wiester H, Shustov G V, Yang D. Can. J. Chem. 1994, 72: 506.
[3] Wu G. Raman spectroscopy. Singapore: World Scientific, 2017.

13 Molecular highly excited vibration

13.1 Introduction

Intermode couplings and nonlinear effects are eminent in molecular highly excited vibration. Then the normal mode picture based on harmonic approximation is no longer appropriate. What is its physical nature is an important topic. With the advance of laser technology, now more of its information has been collected, which allows us to build up its physical picture. The probing to the essentials of molecular highly excited vibration can be the future goal of molecular spectroscopy.

Molecular highly excited vibration, due to its high excitation, is close to the classical domain, though its levels are quantized. Then, a polyatomic molecule is a many-body system with strong nonlinear couplings. Hence, it is expected that the properties and phenomena of the classical nonlinear system like chaos can emerge therein. With this understanding, we may approach this topic by an alternative way, the classical nonlinear dynamics. Spectroscopically, we will find that its classical nonlinear characters will be expressed in the level spacing. Then the core issue is: how to extract out the nonlinear properties including chaos from the quantized level spacing which is the experimentally observable quantity.

This chapter introduces the basic concepts of nonlinear dynamics and their connection to the highly excited vibrational spectrum.

13.2 Morse oscillator

Morse oscillator is an adequate model that can reproduce the bond dissociation. This model realizes the basic characters of a nonlinear system: bound region, dissociation region and the separatrix, which divides these two dynamical regions. The potential of Morse oscillator is

$$D[1-\exp(-ar)]^2$$

Its eigenenergy is

$$E=\omega\left(n+\tfrac{1}{2}\right)+X\left(n+\tfrac{1}{2}\right)^2$$

where n is the quantum number. It is the energy of the simple harmonic oscillator with a second-order correction term. We note that close to the separatrix, the level spacing becomes smaller, as shown in Fig. 13.1.

https://doi.org/10.1515/9783110625097-013

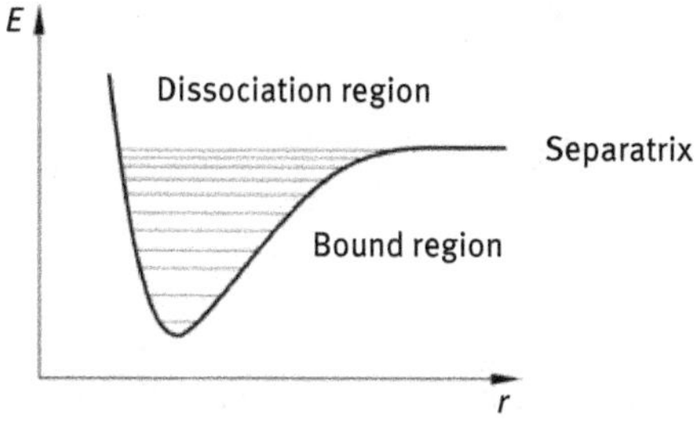

Fig. 13.1: Morse oscillator that shows the basic characters of a nonlinear system: bound region, dissociation region and the separatrix that divides these two dynamical regions. Horizontal lines show the quantized levels. Note that close to the separatrix, the level spacing becomes smaller.

13.3 Pendulum dynamics

The motion of a pendulum is well known. It includes stable and unstable fixed-points. As the swing amplitude is small, it is a simple harmonic motion. When the amplitude is large, it becomes a nonlinear motion. Then its angular velocity ω is dependent on action n. Pendulum dynamics includes bound and unbound (the rotation over the unstable fixed point) regions and their separatrix, similar to the Morse oscillator. Noticeable is that in high dimension, *chaos appears first around the unstable fixed point*, if it does. This is understandable if we consider the situation when a person stands on a big ball (on the unstable "pole"), he has to step forward and backward to keep balance in a random way. His foot trace is nonperiodic and chaotic. This confirms that if chaotic motion appears, it will first appear around the unstable fixed point. Pendulum and Morse oscillator share the common property that in the bound region, level spacing between quantized levels becomes smaller near the separatrix.

In summary, pendulum motion contains the core elements of nonlinear dynamics, as shown in Fig. 13.2.

13.4 Algebraic Hamiltonian

For a simple harmonic oscillator, besides the dynamical variables, the coordinate x and its conjugate momentum p_x, there are the creation and destruction operators, a^+, a that satisfy the commutation relation:

$$[a, a^+] \equiv aa^+ - a^+a = 1$$

This is the so-called second quantization. These two representations are equivalent.

In the second quantization language, the eigenstate of the simple harmonic oscillator can be represented by $|n>$ with integer n, which can be understood as

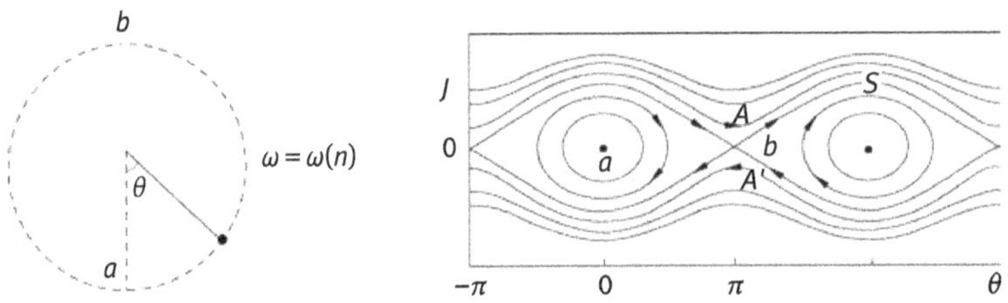

Fig. 13.2: The pendulum motion and its phase diagram. *J* is the action and *θ* is the angle. Curves correspond to the various quantized levels. Arrows denote the directions of motion. *A*, *A′* show the opposite rotations. *S* is the separatrix. *a* and *b* are the stable and unstable fixed points, respectively.

the number of quanta. Each quantum possesses an energy entity of $\hbar\omega_0$ with ω_0 the classical angular frequency. Under the operation of a^+, a, we have the following:

$$a^+ \,|n> = \sqrt{n+1}\,|n+1>$$
$$a\,|n> = \sqrt{n}\,|n-1>$$

The Hamiltonian of the simple harmonic oscillator is

$$H_0 = \hbar\omega_0(\hat{n} + \frac{1}{2})$$

with $\hat{n} = a^+ a$, and

$$\hat{n}\,|n> = n|n>$$

The eigenenergy is $\hbar\omega_0(n + 1/2)$. These compact expressions show the power of second quantization.

For the molecular highly excited system, we will adopt the second quantization algorithm. Such Hamiltonian is called algebraic/spectroscopic Hamiltonian. As the vibrational frequencies of two bonds are close to each other, energy is easier to be transformed between them. This is *resonance*. For example, there can be energy transfer between the two O–H bonds (labeled by *s,t*) in H_2O. This is the 1:1 resonance. Besides, there is the Fermi resonance, or 1:2 resonance between the H–O stretch and the H–O–H bend (labelled by *b*) due to the fact that the bend frequency is half that of the stretch. On the basis of the Morse model, we have for the triatomic system the algebraic Hamiltonian:

$$H = H_0 + H_{st} + H_F$$

$$H_0 = \omega_s(n_s + n_t + 1) + \omega_b(n_b + \tfrac{1}{2}) + X_{ss}[(n_s + \tfrac{1}{2})^2 + (n_t + \tfrac{1}{2})^2]$$
$$+ X_{bb}(n_b + \tfrac{1}{2})^2 + X_{st}(n_s + \tfrac{1}{2})(n_t + \tfrac{1}{2}) + X_{sb}(n_s + n_t + 1)(n_b + \tfrac{1}{2})$$
$$H_{st} = K_{st}(a_s^+ a_t + h.c.)$$
$$H_F = K_{sb}(a_s^+ a_b a_b + a_t^+ a_b a_b + h.c.)$$

where ω, X are the mode frequency and nonlinear coefficients. K is the coupling coefficient.The eigenenergies of this Hamiltonian can be obtained by diagonalizing its matrix constructed by $|n_s, n_t, n_b\rangle$ with n_s, n_t, n_b, the quantum numbers for the two bond stretches and the bend under H_0, provided that the coefficients are known. For this system, although individual n_s, n_t, n_b quantum numbers are destroyed due to couplings, there is still the preserved quantum number $P = n_s + n_t + n_b/2$. The Hamiltonian coefficients can be determined by the fit of the calculated eigenenergies to the experimentally observed mode frequencies. They are [1]: ω_s= 3,890.6 cm^{-1}; ω_t= 3,890.6 cm^{-1}; ω_b= 1,645.2 cm^{-1}; X_{ss}= –82.1 cm^{-1}; X_{tt}= −82.1 cm^{-1}; X_{bb}= −16.2 cm^{-1}; X_{st}= −13.2 cm^{-1}; X_{sb}= −21.0 cm^{-1}; K_{st}= −42.7 cm^{-1}; K_{sb}= −14.5 cm^{-1}.

13.5 Equivalence between resonance and pendulum dynamics

A resonance is equivalent to the pendulum dynamics. We explain this by the 1:1 resonance.

In classical dynamics, Hamilton's equations of motion for action–angle variables (J, θ) are as follows:

$$\partial H/\partial J = \dot{\theta} = \omega\,, \; \partial H/\partial\theta = -\dot{J}$$

The dot above the variable is time derivative.

Suppose the system Hamiltonian is $H_0(J_1, J_2)$. By Hamilton's equations of motion:

$$\omega_1^0 = \partial H_0/\partial J_1, \quad \omega_2^0 = \partial H_0/\partial J_2$$

and $\omega_1^0 \cong \omega_2^0$ or $\omega_1^0 - \omega_2^0 = 0$
Under resonance, the Hamiltonian is

$$H = H_0(J_1, J_2) + C_0(J_1 J_2)\sin(\theta_1 - \theta_2)$$

By the transformation of variables:

$$I_1 = J_1 - J_2, \quad \phi_1 = \theta_1 - \theta_2$$
$$I_2 = J_1 + J_2, \quad \phi_2 = \theta_1 + \theta_2$$

then

$$H = H_0(I_1, I_2) + C_0(I_1 I_2)\sin\phi_1$$

Since H is independent of ϕ_2, $\partial H/\partial\phi_2 = -\dot{I}_2 = 0$, I_2 is conserved. Then,

$$H \cong (\partial H_0/\partial I_1)_0(I_1 - I_1^0) + \tfrac{1}{2}(\partial^2 H_0/\partial I_1{}^2)_0(I_1 - I_1^0)^2 + C_0(I_1^0, I_2^0)\sin\phi_1$$

Since

$$(\partial H_0/\partial I_1)_0 = \frac{\partial H_0}{\partial J_1}\frac{\partial J_1}{\partial I_1} + \frac{\partial H_0}{\partial J_2}\frac{\partial J_2}{\partial I_1} = \omega_1^0 - \omega_2^0 = 0$$

then finally we have

$$H \sim \frac{1}{2}\alpha I_1^2 + C_0 \sin\phi_1 \ (\alpha,\ C_0 \text{ are constant})$$

This is just the Hamiltonian of a pendulum with kinetic energy $\frac{1}{2}\alpha I_1{}^2$ in a potential of $C_0 \sin\phi_1$. Therefore, a resonance is equivalent to the pendulum dynamics. The inference is: a resonance contains the ingredients of nonlinear dynamics like stable and unstable fixed points, separatrix, regular and irregular/chaotic motions.

13.6 A resonance is associated with a constant of motion

From the algebraic Hamiltonian shown above for H_2O (Section 13.4), for the 1:1 resonance, $P = n_s + n_t$ is a constant of motion and so is $P = n_s + n_t + n_b/2$ for the 1:2 Fermi resonance. As both 1:1 and 1:2 resonances exist, $P = n_s + n_t + n_b/2$ is still a constant of motion. These conserved quantum numbers are called polyad numbers. We can employ polyad number to classify the system levels when there is resonance. Since the resonance is equivalent to the pendulum dynamics, the levels associated with a polyad number are just like the quantized levels of a pendulum, for which the level spacing will reach a minimum as they cross the separatrix (Section 13.3). This is called the Dixon dip [2].

13.7 Chaos

Intuitively, chaos means disorder, irregular or random motion. In fact, disorder, irregularity or random depends on our perception. A disorder, irregular or random system may reveal order structure if our perception is critical enough. The number of molecular highly excited vibrational levels is immense. The distribution of these levels seems irregular. However, if they are analyzed by appropriate methods, embedded regular structure will be revealed.

Suppose that initially the deviation between two moving points is $\Delta(0)$ and at time t, the deviation is $\Delta(t)$. If

$$\Delta(t) \approx \Delta(0)e^{ht}$$

and $h > 0$, then the trajectories are very initial point dependent. h, called the *Lyapunov exponent*, is a parameter for describing the *degree* of chaos of the trajectories. Now, the accepted definition of chaos is: a trajectory is chaotic if it follows a deterministic equation of motion and its Lyapunov exponent is larger than 0. For an N-dimensional system, there are N Lyapunov exponents. Often, we need only the largest Lyapunov exponent since it the major factor determining the degree of chaos.

13.8 Heisenberg correspondence

Heisenberg recognized that creation and destruction operators can be associated with the classical action–angle variables [3], that is,

$$a^+ \approx \sqrt{n}\,\mathrm{e}^{i\phi},\; a \approx \sqrt{n}\,\mathrm{e}^{-i\phi}$$

(n, ϕ) coordinates can be expressed by the generalized coordinates (q, p)such as

$$q = \sqrt{2n} \quad \cos\phi, \quad p = -\sqrt{2n} \quad \sin\phi$$

These coordinates in fact are the coordinates for the coset space in Lie algebra/ group.

For example, for the 1:1 resonance in Section 13.4, its second quantization representation can be cast first into that by (n, ϕ) and then to (q, p) as

$$a_s^+ a_t + a_t^+ a_s \sim q_s q_t + p_s p_t$$

Since for this interaction, the total action $n_s + n_t$ is conserved, another choice is

$$a_s^+ \sim \sqrt{n_s}, \quad a_t^+ \sim \sqrt{n_t} e^{i\phi_t}$$

where ϕ_t is the phase difference between the two actions of s and t. Then we have the following:

$$a_s^+ a_t + a_t^+ a_s \sim \sqrt{2n_s} q_t$$

In summary, for a given polyad number P, we have:

$n_t = (q_t^2 + p_t^2)/2$, $n_b = (q_b^2 + p_b^2)/2$, $n_s = P - (n_t + n_b/2)$. Coupling terms are K_{st} $(2n_s)^{\frac{1}{2}} q_t$ (for 1:1 resonance), $K_{DD} n_s (q_t^2 - p_t^2)$ (for $a_s^+ a_s^+ a_t a_t + h.c.$ the 2:2 resonance or-Darling–Dennison coupling) and $K_{sb}\{\sqrt{n_s}(q_b^2 - p_b^2) + [q_t(q_b^2 - p_b^2) + 2p_t q_b p_b]/\sqrt{2}\}$ (for Fermi resonance).

Then, the Hamiltonian is cast in terms of (q_t, p_t, q_b, p_b) variables as $H = H(q_t, p_t, q_b, p_b)$. Hamilton's equations of motion are

$$\partial H/\partial q_\alpha = -dp_\alpha/dt$$

$$\partial H/\partial p_\alpha = dq_\alpha/dt \quad (\alpha = t, b)$$

The motion trajectory forms a three-dimensional subspace in a four-dimensional (coset) space since the system energy is conserved. Given the initial $(q_t, p_t, q_b, p_b)_0$, the motion trajectory moves by Hamilton's equations.

As an example, the frequencies of D–C and C–N of DCN are 2,681.4 and 1,948.9 cm^{-1}, respectively . Their ratio is 1.37. For this system, there can be 1:1 and 2:3 resonances. Their polyad numbers are $P_1 = n_s + n_t$ and $P_2 = n_s/2 + n_t/3$. Since these two resonances are coexistent, these two polyad numbers are only approximate. Although not exact, they are still useful. Figure 13.3 shows the levels classified by these two polyad numbers [4].

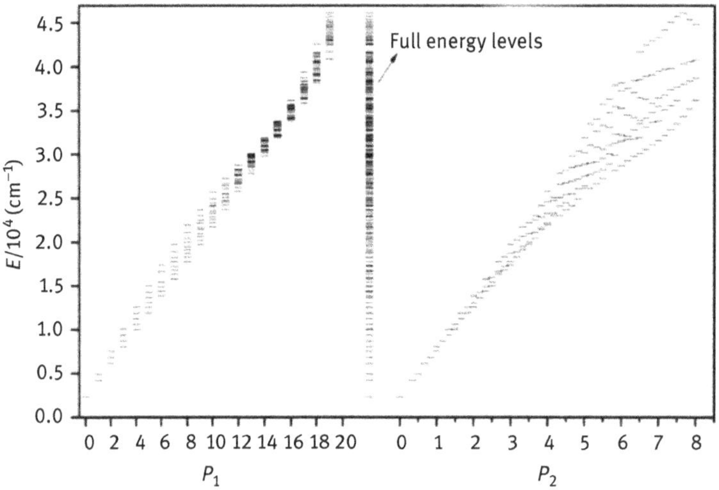

Fig. 13.3: The DCN levels due to D–C and C–N stretches below 45,000 cm^{-1} classified by $P_1 = n_s + n_t$ and $P_2 = n_s/2 + n_t/3$.

The levels classified by the polyad numbers, P_1 and P_2, can be employed to analyze their Dixon dips. The result shows that for $P_1 < 7$ (energy < 20,000 cm^{-1}), the nearest neighboring level spacings show distinct dips as shown in Fig. 13.4(a). For higher levels, ($P_1 > 7$), no dips are distinct. Then, the 1:1 resonance is perturbed by the 2:3 resonance. For $P_2 < 5$ (energy < 30,000 cm^{-1}), dips are not obvious. However, for larger P_2 (energy is around 45,000 cm^{-1}), dips appear again, as

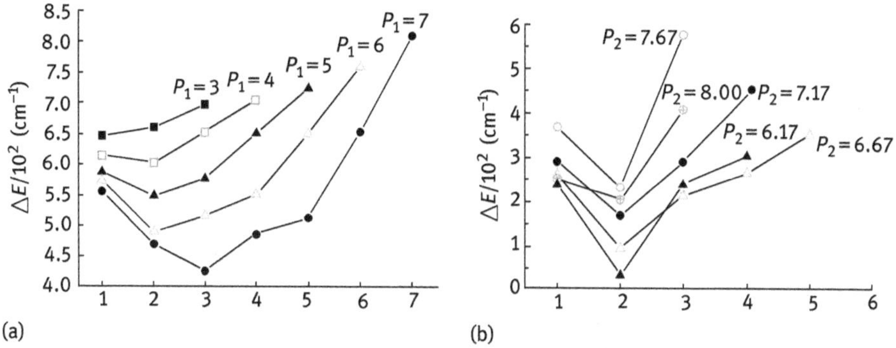

Fig. 13.4: The nearest neighboring level spacings for DCN (a) $P_1 < 7$ and (b) $P_2 > 5$.

shown in Fig. 13.4(b). This contrast convinces us that for the levels between 20,000 and 30,000 cm^{-1}, the two resonances interfere with each other seriously that their separatrices are distorted and Dixon dips are not obvious. In this energy region, 1:1 and 2:3 resonances overlap. The following analysis demonstrates that the region, in which there is resonance overlap, is most dynamically chaotic.

13.9 Resonance overlap leads to chaos

Chirikov has proposed a conjecture [5] that resonance overlap leads to chaos. As demonstrated earlier, for DCN levels in between 20,000 and 30,000 cm^{-1}, 1:1 and 2:3 resonances overlap. We will test Chirikov's viewpoint by this case. First, the algebraic Hamiltonian for this system is cast to the dynamical variables in the coset space. For each state, we randomly choose 200 initial points to calculate its average (largest) Lyapunov exponent, $\langle \lambda \rangle$. The results are shown in Fig. 13.5 [4]. Obviously, as level energy reaches 10,000 cm^{-1}, state dynamics starts showing chaos. In this energy region, the 1:1 resonance is the main dynamics. As energy reaches 25,000 cm^{-1}, dynamical chaos reaches its highest degree ($\langle \lambda \rangle$ is maximal). In between 20,000 and 30,000 cm^{-1}, both 1:1 and 2:3 resonances operate. As pointed out by Chirikov, their overlap leads to chaos. As level energies are larger than 30,000 cm^{-1}, 2:3 resonance plays the dominant role. Then the overlap and Lyapunov exponent decrease slightly. For higher levels, the degree of chaos decreases instead.

Up to this point, we know, besides that chaos can emerge around the unstable fixed point, resonance overlap may also lead to chaos. The former chaos is local, while the latter one may lead to the global effect.

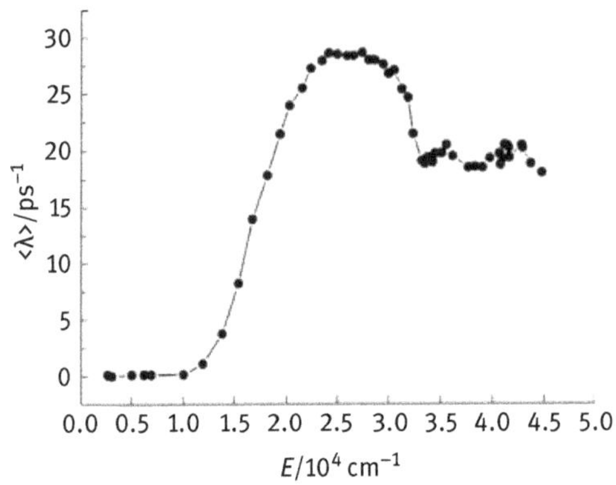

Fig. 13.5: The average largest Lyapunov exponent as a function of energy for the DCN system.

13.10 Dynamical potential

1. Fixed point

The dynamical potential for a system, whose Hamiltonian is $H(P, q_\alpha, p_\alpha)$ with P given, is the *effective environment* in which the q_α coordinate experiences [6]. This is achieved by calculating the maximal and minimal energies by varying p_α for each q_α under the condition that actions are nonnegative. The points composed of these maximal and minimal energies as a function of q_α form a closed curve in which all the quantized levels possessing the same P will reside. The points on the dynamical potential curve correspond to $\partial H/\partial p_\alpha = 0$. Among them, those satisfy $\partial H/\partial q_\alpha = 0$ are the fixed points for the system corresponding to P.

In HCP, the internal rotation of H around C–P can be treated as the 1:2 resonance between rotation (label 2) and C–P stretch (label 3). The dynamical potential corresponding to $P = n_2 + 2n_3 = 22$ is shown in Fig. 13.6. The 12 levels embedded therein are also shown. In the figure, stable and unstable fixed points are $[B]$, $[r]$, $[SN]$ and $[\overline{SN}]$. We note that quantized levels are governed by the dynamical potential. Dynamical potential is *defined* by the fixed points. In other words, as fixed points are given, dynamical potential is *almost* defined, so are the characters of the quantized levels. There is close relation between the quantized levels and the *classical fixed points*. The quantized levels just reside in the quantal *environments* defined by the fixed points.

The trajectories ($p_\alpha = p_\alpha(q_\alpha)$, $(\alpha = 2, 3)$)for these levels can be obtained from the equation: eigenenergy = $H(P, q_\alpha, p_\alpha)$. The trajectories are closed. Although they are classical, their characters are consistent with those by the wave functions through quantal algorithm [6, 7].

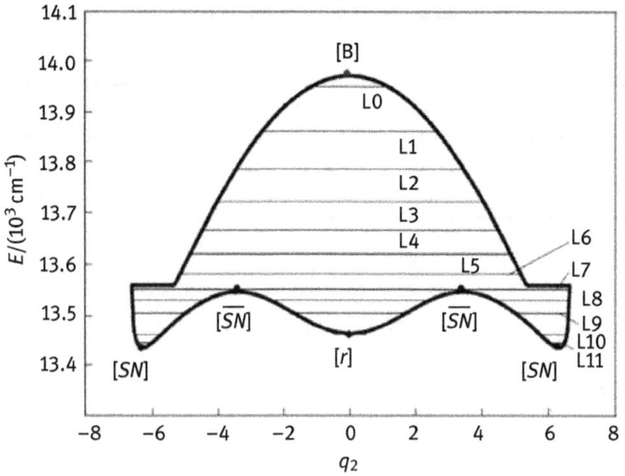

Fig. 13.6: The dynamical potential corresponding to $P=22$ of HCP system. $[B]$, $[r]$, $[SN]$ and $[\overline{SN}]$ are the stable and unstable fixed points. Horizontal segments are the quantized levels.

2. Quantal environments

The action integral (the classical analogue of quantum number) for the trajectory is

$$1/2\pi \oint p_\alpha dq_\alpha \; (\alpha = 2, 3)$$

The results from (q_2, p_2) space are given in Tab. 13.1. These levels can be categorized into three classes: L0 to L7, L8 to L10 and L11. The action integral differences between neighboring levels are quite the same, around 2. (This value could be 1 if the polyad number is redefined as $n_2/2 + n_3$.) The action integral from L0 to L7 increases. This is understandable since these levels stay in an *inverse* harmonic-like potential (Fig. 13.6) that for the lower levels, their action integrals (quantum numbers) are larger. For the levels from L8 to L10, their action integrals decrease since they lie in a *regular* potential. L11 is, by itself, a unique state. We recognize that for the quantized levels residing in a quantal environment, the difference of the action integrals/quantum numbers of the neighboring levels is a constant. Hence, we say that L0 to L7, L8 to L10 and L11 are in three different quantal *environments*, respectively. Such classification is consistent to that by the wave functions: L0–L7, L8–L10 and L11 possess three different patterns for their wave functions [7]. In the table, there are the results in the (q_3, p_3) space. The results are consistent with those in (q_2, p_2) space. (Of course, the classification can be as L0–L10 and L11.)

3. Localized mode

For each level, from its trajectory, the relation $n_\alpha = n_\alpha(q_\alpha)$ can be determined. By the integration against q_α, the action percentages of the internal rotation of H around C–P and the C–P stretch are obtained.

Tab. 13.1: The action integrals and their differences of the 12 quantal levels corresponding to $P = 22$ for HCP.

Level label	Action integral in (q_2, p_2) space	Difference of the action integrals of the neighboring levels	Action integral in (q_3, p_3) space	Difference of the action integrals of the neighboring levels
L0	0.5	/	0.2	/
L1	2.5	2.0	1.3	1.0
L2	4.5	2.0	2.3	1.0
L3	6.6	2.1	3.3	1.0
L4	8.7	2.1	4.3	1.0
L5	10.8	2.1	5.4	1.1
L6	12.9	2.1	6.5	1.1
L7	15.4	2.5	7.6	1.1
L8	4.6	/	8.7	/
L9	2.8	1.8	9.6	0.9
L10	0.6	2.2	10.8	1.2
L11	0.2	/	0.1	/

We found that the lowest level in the dynamical potential corresponding to a polyad number possesses the highest percentage of localized rotational/ bend energy. This level, which is near the fixed point [*SN*], is the most possible candidate for the transition to HPC through HCP bending motion. For convenience, we call it [*SN*] mode. For the other higher levels, due to intramolecular vibrational relaxation (IVR), their energy is distributed among the bond motions so that they are unable to realize the transition. Even for the highest level, its action percentage of HCP bend is smaller than that of the [*SN*] mode as shown in Fig. 13.7. Moreover, its energy on the bend coordinate is less than that of the [*SN*] mode. From Fig. 13.7, we note that for the highest level, the action percentage of $n_2/2$ drops as P is larger. This is due to IVR. On the contrary, for the [*SN*] mode, IVR does not enhance as P is larger. Its localized character enhances as its level energy increases.

4. Dynamical potential similarity

In DCP, D–C and C–P stretches possess 1: 2 resonance and have little coupling with the bend motion. Its dynamical potentials share similarity with those of HCP as shown in Fig. 13.8. By their comparison, we know that in DCP, there is localized mode close to the fixed point $[R_1]$. It is the highest level. There are two quantal environments for DCP. This demonstrates that by knowing the dynamical potential of a system, the level properties of other systems that share similar potentials can be

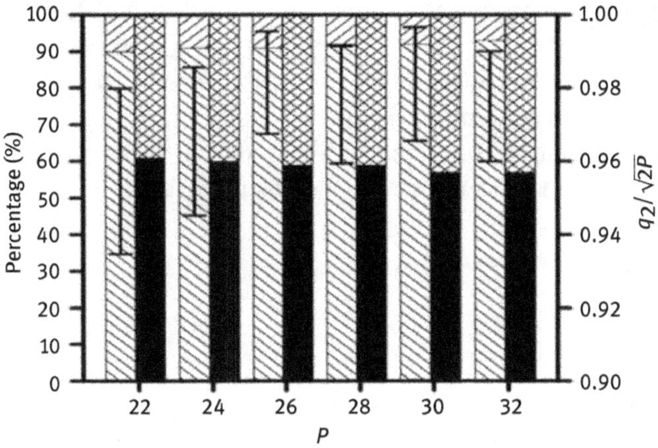

Fig. 13.7: The action percentages of $n_2/2$ (left-up shadow) and n_3 (right-up shadow) for the [SN] mode as the functions of P. Bar I shows the allowed range of q_2 coordinate. This shows the localization of the [SN] mode. Black and cross bars are for $n_2/2$ and n_3, respectively, of the highest levels for various P, for which IVR is evident.

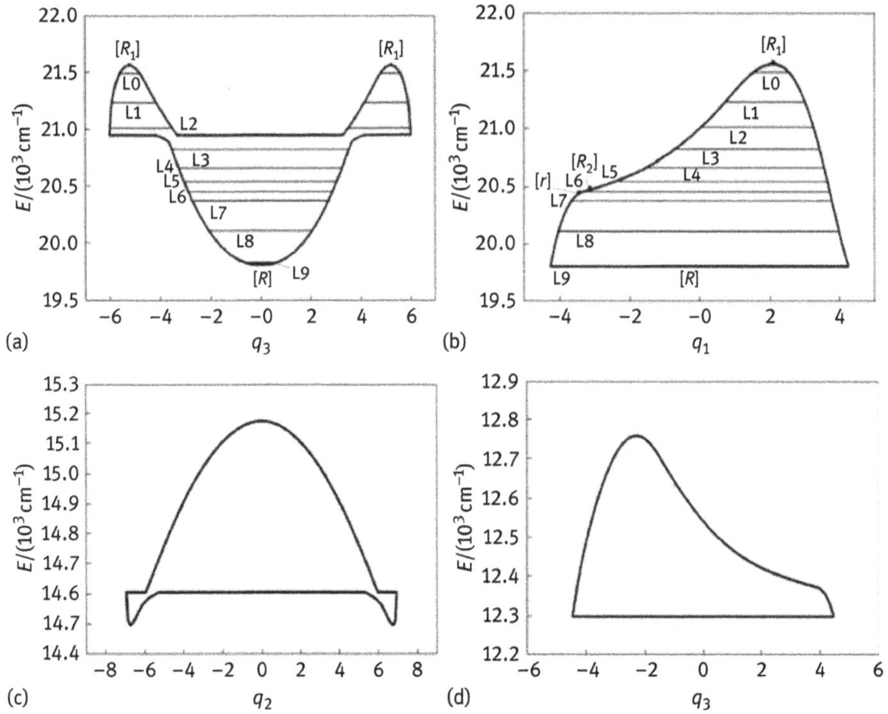

Fig. 13.8: The dynamical potentials (a) and (b) of DCP with $P = 18$ as contrasted with those of HCP in q_2 (P= 24) (c) and q_3 (P= 20) coordinates (d). (a) and (c) are similar if inverted. (b) and (d) are similar if $-q_3$ is substituted for q_3. $[R_1]$, $[R_2]$, $[R]$ and $[r]$ are fixed points. The horizontal lines show the quantal levels.

elucidated at hand. We stress that dynamical potential is classical and level is quantal.

These results show that different molecules are not so unrelated as far as their state dynamics or spectroscopy is concerned. The fact is that they may share common characters, which can be out of expectation at a first glance. These common characters are not so easy to find out simply by the algorithm based on wave function.

13.11 Conclusion

The method presented in this chapter combines the ideas from second quantization, Heisenberg's correspondence, classical dynamics – Hamilton's equations of motion, pendulum dynamics, nonlinear dynamical concept – Lyapunov exponent, chaos and so on. The data are from the experimental observation – the level spacing between the quantized levels [8].

Molecular highly excited vibration, although its dynamics is quantized, still preserves many classical remnants as shown in Fig. 13.9. Classical dynamics, nonlinear dynamics and their related concepts are not only useful but also necessary for the exploration. We expect that embedded in a quantized system, there are still many classical properties. Nonlinear and classical algorithms are useful to understand the quantal system. They enable us to obtain the *global* properties of the quantal system. These would be difficult to obtain by the algorithm based on wave function. In addition, some properties masked by wave function algorithm can be exposed thereby. In this field, the quantal algorithm by solving Schrödinger equation is not the only choice. Other classical nonlinear algorithms are possible.

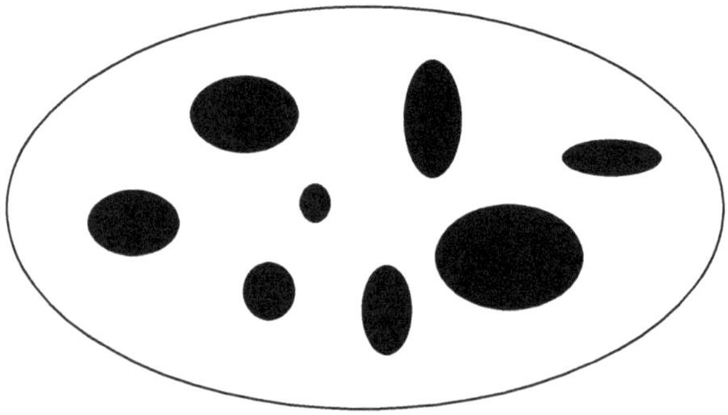

Fig. 13.9: In a quantal system, there are many embedded classical information and properties. The black portions symbolize the classical realms.

References

[1] Iachello F, Oss S. J. Mol. Spectrosc. 1990, 142: 85
[2] Dixon RN. Trans. Farad. Soc. 1964, 60: 1363
[3] Heisenberg W. Z. Physik 1925, 33: 879
[4] Wang H, Wang P, Wu G. Chem. Phys. Lett. 2004, 399: 78
[5] Chirikov BV. Phys. Rep. 1979, 52: 263
[6] Fang C, Wu G. J. Mol. Struct.: Theochem 2009, 910: 141
[7] Joyeux M, Sugny D, Tyng V, Kellman M, Ishikawa H, Field R, Beck C, Schinke R. J. Chem. Phys. 2000, 112: 4162
[8] Wu G. Molecular vibrations: An algebraic and nonlinear approach. Singapore: World Scientific, 2018.

Answers to exercises

Chapter 1

1.1 From the completeness $\sum_{\upsilon} |\upsilon\rangle\langle\upsilon| = I$,

we have $\langle\upsilon|x^n|\upsilon'\rangle = \sum_{\upsilon_1}\sum_{\upsilon_2}\cdots\sum_{\upsilon_{n-1}} \langle\upsilon|x|\upsilon_1\rangle\langle\upsilon_1|x|\upsilon_2\rangle\cdots\langle\upsilon_{n-1}|x\upsilon'\rangle$

We need to consider those nonzero terms with $\upsilon_i = \upsilon_j \pm 1$.

1.2 Substitute $g(\omega) = \delta(\omega - \omega_0)$ into $P_m(t)$ and with the property of δ

$$\int f(\omega)\delta(\omega - \omega_0)d\omega = f(\omega_0)$$

Then we have $P_m(t) \sim t^2$.

1.3 Since $e^{ikz} = 1 + ikz + \cdots$, each term is smaller than its previous one by the factor of kz, and the corresponding radiation strength is smaller by $(kz)^2 \sim \left(\frac{2\pi}{\lambda}a\right)^2$. For $\lambda \sim 5,000$ Å, a is the atomic dimension, then

$$\left(\frac{2\pi}{\lambda}a\right)^2 \sim 10^{-6}$$

1.4 Consider the integration by parts

$$\int_{-\infty}^{\infty} \psi_k x \frac{d^2\psi_m^*}{dx^2}dx = \psi_k x \frac{d\psi_m^*}{dx}\Big|_{-\infty}^{\infty} - \int \frac{d}{dx}[\psi_k x]\frac{d\psi_m^*}{dx}dx$$

and note that $d(\psi_m^*\psi_k) = \psi_k d\psi_m^* + \psi_m^* d\psi_k$.

1.5 By Bohr's atomic model

$$\nu_{nm} \approx Z^2\left(\frac{1}{n^2} - \frac{1}{m^2}\right), \text{ i.e., } \nu_{nm} \approx Z^2$$

The electron energy $E \approx \frac{Z}{r}$ or

$\approx \frac{Z}{E}$, and $E \approx \frac{Z^2}{n^2}$, then $r \approx \frac{Z}{Z^2} = \frac{1}{Z}$ and $|X_{nm}| \approx r \approx \frac{1}{Z}$. Moreover, $A_{nm} \approx \nu^3|X_{nm}|^2$ $\approx Z^6/Z^2 \approx Z^4$.

Hence, the lifetime of the excited He^+ ($Z = 2$), $\tau \sim A_{nm}^{-1}$ is 1/16 of H.

1.6 Consider the Fourier transform between $I(\omega)$ and $\langle\mu(0)\mu(t)\rangle$.

1.7 Substitute values into

$$\Delta\omega_h^D = 2\omega_0\left[2\ln 2\frac{kT}{Mc^2}\right]^{1/2}$$

For $T = 300\text{K}$, $M = 20$, $\lambda = 6,000$ Å, the Doppler's width is 1.66×10^{-2} Å. This is two orders larger than the natural width.

https://doi.org/10.1515/9783110625097-014

1.8 Spectral width γ is the sum of those of the initial and final states. E_1 is the ground state, and its energy width is 0 (since its lifetime is infinite). Hence,

$$\gamma_{31} = \Delta E_3 = \tau_3^{-1}$$
$$\gamma_{32} = \Delta E_3 + \Delta E_2 = \tau_3^{-1} + \tau_2^{-1}$$

τ^{-1} is the sum of the probabilities by spontaneous emissions. Then

$$\tau_3^{-1} = A_{31} + A_{32}$$
$$\tau_2^{-1} = A_{21}$$

and

$$\gamma_{31} = \tau_3^{-1} = A_{32} + A_{31}$$
$$\gamma_{32} = \tau_3^{-1} + \tau_2^{-1} = A_{32} + A_{31} + A_{21}$$

Chapter 2

2.1 (1) If we require that as $R \to \infty, E = E_k = E_p = 0$, E, E_k and E_p have to be added with −De, De and −2De, respectively. Their relation with R is as shown below. (suppose $\beta = 1, R_e = 1.40$ a.u.)

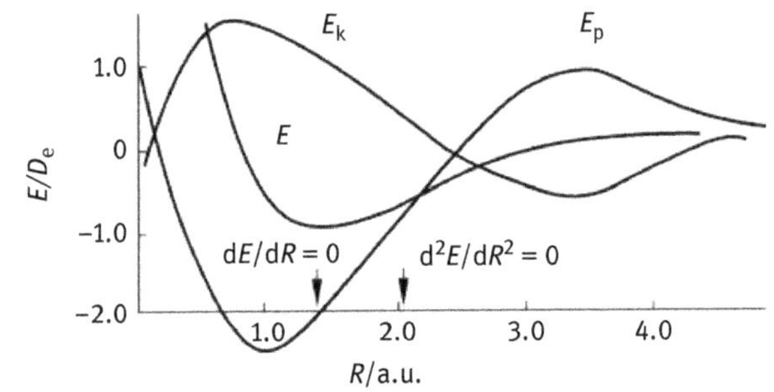

(2) Differentiate E twice with respect to R.

(3) As $R_e < R < R_e + \frac{1}{\beta}\ln 2$, the decrease of E_p is faster than the increase of E_k. As $R > R_e + \frac{1}{\beta}\ln 2$, the decrease of E_k is faster than the increase of E_p.

2.2 By $dN_J/dJ = 0$, then $2J + 1 = [2kT/B]^{1/2}$.

2.3 The spin of ^{16}O is 0. It is a boson. $^{16}O_2$ can only have spin of 0. Its spin wave function is symmetric. Its vibrational wave function is also symmetric. Then, J can only be odd such that the overall wave function is symmetric.

^{17}O is a fermion. The spin of $^{17}O_2$ can be 0, 1, 2, 3, 4, 5. As the spin is 0, 2, 4, the spin wave function is antisymmetric while for spin 1, 3, 5, it is symmetric. Hence, for the former case, J can only be odd. For the latter case, J can only be even.

2.4 $E_J = J(J+1)B_J$

$$\bar{\nu} = E_{J'} - E_{J''} = J'(J'+1)B' - J''(J''+1)B''$$

For P branch $J' = J'' - 1$,

$$\bar{\nu} = (B' - B'')J''^2 - (B' + B'')J''$$

$J'' > J'$, then $B' > B''$
Hence, for larger J'', the spectral line spacing is larger.
For R branch $J' = J'' + 1$

$$\bar{\nu} = (B' - B'')(J''+1)^2 + (B' + B'')(J''+1)$$

$J' > J''$, then $B' < B''$
Hence, for larger J'', the spectral line spacing is smaller.

2.5

$$\Delta E_V = \omega_e - 2(V+1)\omega_e x_e$$

Let $\Delta E_V = y$, $\omega_e(1-2x_e) = a$, $-2\omega_e x_e = m$, $V = x$
Then $y = a + mx$. By linear fitting, a, m and ω_e, x_e can be obtained.
Or by $V_{\max} = \frac{1}{2x_e} - 1$, we can obtain $\omega_e \sim 4,404.51\ \text{cm}^{-1}$, $x_e \sim 0.0266$, $V_{\max} \sim 18$.
Above the $V_{\max}$ is obtained by linear extrapolation. In fact, ΔE_V and V are not of linear relation. The real $V_{\max}$ is when $\Delta E_V = 0$, and then $V_{\max} \sim 15$.

Chapter 3

3.1 (1) $P_{S_1 i} = \frac{\partial T}{\partial \dot{S}_{1i}} = \sum_k \frac{\partial T}{\partial \dot{S}_{2k}} \frac{\partial \dot{S}_{2k}}{\partial \dot{S}_{1i}} = \sum_k P_{S_2 k} A_{ki}$

that is, $\boldsymbol{P}_{S_1} = \boldsymbol{A}^{\mathrm{T}} \boldsymbol{P}_{S_2}$

(2) Since

$$T_{S_1} = T_{S_2}$$
$$\boldsymbol{P}_{S_1}^{\mathrm{T}} \boldsymbol{G}_{S_1} \boldsymbol{P}_{S_1} = \boldsymbol{P}_{S_2}^{\mathrm{T}} \boldsymbol{G}_{S_2} \boldsymbol{P}_{S_2}$$

also $\boldsymbol{P}_{S_1} = \boldsymbol{A}^{\mathrm{T}} \boldsymbol{P}_{S_2}$, $\boldsymbol{P}_{S_1}^{\mathrm{T}} = \boldsymbol{P}_{S_2}^{\mathrm{T}} \boldsymbol{A}$
then, $\boldsymbol{A} \boldsymbol{G}_{S_1} \boldsymbol{A}^{\mathrm{T}} = \boldsymbol{G}_{S_2}$
Since $V_{S_1} = V_{S_2}$, $\boldsymbol{S}_1^{\mathrm{T}} \boldsymbol{F}_{S_1} \boldsymbol{S}_1 = \boldsymbol{S}_2^{\mathrm{T}} \boldsymbol{F}_{S_2} \boldsymbol{S}_2$, $\boldsymbol{S}_2 = \boldsymbol{A} \boldsymbol{S}_1$ or $\boldsymbol{S}_2^{\mathrm{T}} = \boldsymbol{S}_1^{\mathrm{T}} \boldsymbol{A}^{\mathrm{T}}$, then $\boldsymbol{F}_{S_2} = (\boldsymbol{A}^{\mathrm{T}})^{-1} \boldsymbol{F}_{S_1} \boldsymbol{A}^{-1}$
From(2) $\boldsymbol{G}_{S_2} \boldsymbol{F}_{S_2} = \boldsymbol{A} \boldsymbol{G}_{S_1} \boldsymbol{A}^{\mathrm{T}} (\boldsymbol{A}^{\mathrm{T}})^{-1} \boldsymbol{F}_{S_1} \boldsymbol{A}^{-1} = \boldsymbol{A} \boldsymbol{G}_{S_1} \boldsymbol{F}_{S_1} \boldsymbol{A}^{-1}$
again, $\boldsymbol{L}_{S_2}{}^{-1} \boldsymbol{G}_{S_2} \boldsymbol{F}_{S_2} \boldsymbol{L}_{S_2} = \boldsymbol{\Lambda} = \boldsymbol{L}_{S_1}{}^{-1} \boldsymbol{G}_{S_1} \boldsymbol{F}_{S_1} \boldsymbol{L}_{S_1} = \boldsymbol{L}_{S_2}{}^{-1} \boldsymbol{A} \boldsymbol{G}_{S_1} \boldsymbol{F}_{S_1} \boldsymbol{A}^{-1} \boldsymbol{L}_{S_2}$
Then $\boldsymbol{L}_{S_2} = \boldsymbol{A} \boldsymbol{L}_{S_1}$

(3) If $\boldsymbol{G}_{S_1} = \boldsymbol{G}_{S_2} = I$, then $\boldsymbol{A}\boldsymbol{G}_{S_1}\boldsymbol{A}^{\mathrm{T}} = \boldsymbol{G}_{S_2}, \boldsymbol{A}\boldsymbol{A}^{\mathrm{T}} = I$, that is, $\boldsymbol{A}^{\mathrm{T}} = \boldsymbol{A}^{-1}$.
In Section 3.2, $\boldsymbol{L}^{-1}$ satisfies this relation. Hence, $\left(\boldsymbol{L}^{-1}\right)_{ki} = L_{ik}$

3.2 Consider $H_0\psi_1 = \varepsilon_1\psi_1,\ H_0\psi_2 = \varepsilon_2\psi_2,\ \varepsilon_1 \cong \varepsilon_2$
Let $\psi = a\psi_1 + b\psi_2, H = H_0 + H', \mathrm{H}\psi = \varepsilon\psi$
Then $\left(H'_{11} + \varepsilon_1 - \varepsilon\right)a + H'_{12}b = 0$

$$\left(H'_{22} + \varepsilon_2 - \varepsilon\right)b + H'_{12}a = 0$$

leads to

$$\varepsilon_{\pm} = \frac{E_1 + E_2}{2} \pm \sqrt{H'_{12} + \left(\frac{E_1 + E_2}{2}\right)^2}$$

Here, $H'_{ij} = \langle\psi_i|H'|\psi_j\rangle,\ E_1 = \varepsilon_1 + H'_{11},\ E_2 = \varepsilon_2 + H'_{22}$.

3.3 (1) Since μ and r are in proportion, then
$\partial\mu_1/\partial r_1 = \mu_1/\rho$, also $d\mu_1 \cos\theta = d\mu_Z$, $dr_1 = 2/\sqrt{2}dS_4$; hence, $\frac{\partial\mu_1}{\partial r_1} = \frac{1}{\sqrt{2}\cos\theta}\frac{\partial\mu_Z}{\partial S_4}$.
(2) From (1), we have

$$d\mu_1/dr_1 \sin\theta\sqrt{2} = \partial\mu_Z/\partial S_4 \tan\theta$$
$$= d\mu_Y/\tfrac{1}{\sqrt{2}}dr_1 = \partial\mu_Y/\partial S_1$$

(3) $\frac{\partial\mu_Z}{\partial S_5} = \frac{\partial\left[\mu_1\cos(\theta + (\beta/2))\right]}{\rho\partial\beta/2} = \frac{-\mu_1}{\rho}\sin\left(\theta + \frac{\beta}{2}\right)$
(4) From (1), we have $\frac{\partial\mu_Z}{\partial S_4} = \sqrt{2}\cos\theta\frac{\partial\mu_1}{\partial r_1} = \sqrt{2}\cos\theta\frac{\mu_1}{\rho}$.

From (3), we have $\frac{\partial\mu_Z}{\partial S_5} = -\frac{\mu_1}{\rho}\sin\left(\theta + \frac{\beta}{2}\right)$.

Hence, $\frac{\partial S_5}{\partial S_4} = -\frac{\cos\theta\sqrt{2}}{\sin(\theta + \beta/2)}$.

(5) $\frac{\partial\mu_Z}{\partial S_4} = \frac{\partial\mu_Z}{\partial S_5}\frac{\partial S_5}{\partial S_4} = \frac{\partial\mu_Z}{\partial S_5}\frac{-\cos\theta\sqrt{2}}{\sin(\theta + \beta/2)}$.

Chapter 5

5.1 Consider $S_n^n = C_n^n\sigma_h^n = \sigma_h^n$, as n is odd, $\sigma_h^n = \sigma_h$; hence, σ_h exists. Also, $S_n\sigma_h = C_n$; hence, C_n exists.

5.2 Consider two finite cyclic groups

$$G_1 = \left\{a^0, a^1, a^2, \ldots, a^n\right\}$$
$$G_2 = \left\{b^0, b^1, b^2, \ldots, b^n\right\}$$

There can be a correspondence $a^k \leftrightarrow b^k$; hence, G_1, G_2 are isomorphic.

5.3 Choose an element a, and have $a^0, a^1, a^2, \ldots, a^m = a^k (k < m)$. If $m > 5$, then the rank is more than 5. This is contradictory. If $m < 5$, then there is a subgroup. This is contradictory. Conclusion: the group is cyclic.

5.4 For abelian group, $xax^{-1} = axx^{-1} = a$; hence, each element forms a class.

5.5 Left and right cosets are $\{E, C_3, C_3^2\}$, $\{\sigma_v^{(1)}, \sigma_v^{(2)}, \sigma_v^{(3)}\}$.

Three classes are $\{E,\}\{C_3, C_3^2\}$, $\{\sigma_v^{(1)}, \sigma_v^{(2)}, \sigma_v^{(3)}\}$.

5.6 Refer to the elementary textbook on group theory.

5.7 K_i is coset. Multiplication of cosets can be defined by that among the elements in the cosets. The point is to prove that the product of cosets is still a coset.

5.8 $G = H \cup_{g_1} H \cup_{g_2} H \cup \cdots$

where H is an invariant subgroup. Consider the map of the element $g_{i,k}$ of g_iH to g_iH and note that

$$g_{i,k}g_{j,l} \to g_iHg_jH = g_ig_jH$$

This proves the theorem.

5.9 S_6, D_{3d}, D_{5h}, O_h.

5.10 (1)C_{3h}; (2) C_{2h}; (3)D_{6h}; (4) D_{3d}; (5) T_d; (6) D_{2h}.

5.11 T_r denotes the trace. Prove that $T_r(\boldsymbol{AB}) = T_r(\boldsymbol{BA})$. Then, $T_r(\boldsymbol{A}) = T_r(\mathbf{S}^{-1}\mathbf{AS})$.

5.12 From

$$\sum_i \chi^{(i)}(C_k)^* \chi^{(i)}(C_l) = \frac{g}{N_k}\delta_{kl}$$

Let $C_k = C_l = E$, $N_k = 1$, $\chi^{(i)}(E) = d_i$, then $\sum_i d_i^2 = g$.

5.13 Write down the Coulomb potential and make the following transformations: $y_i \to z_i$, $z_i \to -y_i$ and $y_i \to -y_i$, $z_i \to -z_i$.

5.14 Molecular point group is C_{2h}. The symmetries of normal modes are:

$$\Gamma^{\text{vib}} = 5A_g + B_g + 2A_u + 4B_u$$

A_g, B_g are Raman active, and A_u, B_u are i.r. active. The in-plane coordinates are symmetric under σ_h. The out-of-plane coordinates are antisymmetric under σ_h. Hence, A_g, B_u are in-plane modes, and B_g, A_u are out-of-plane modes. As the symmetry is reduced to C_2, A_g, A_u transform to A symmetry, B_g, B_u to B symmetry. Then they are both i.r. and Raman active.

5.15 Choose the internal coordinates:

$$\begin{gathered} r_1(\mathrm{H-C}),\ r_2(\mathrm{Cl-C}), \\ r_3(\mathrm{C=C}),\ r_4(\mathrm{C-Cl}), \\ r_5(\mathrm{C-H}),\ \theta_1(\mathrm{H-C=C}), \\ \theta_2(\mathrm{Cl-C}=C),\ \theta_3(C=\mathrm{C-Cl}), \\ \theta_4(\mathrm{C=C-H}),\ \delta_1(\mathrm{Hout-of-plane}) \\ \delta_2(\mathrm{C=C-Clout-of-plane}),\ \tau(\mathrm{C=Ctorsion}) \end{gathered}$$

From C_{2h} character table and the following projection operators:

$$P_{A_g} = N(E + C_2 + i + \sigma_h),$$
$$P_{B_g} = N(E - C_2 + i - \sigma_h),$$
$$P_{A_u} = N(E + C_2 - i - \sigma_h),$$
$$P_{B_u} = N(E - C_2 - i + \sigma_h),$$

obtain the symmetry coordinates:

Symmetry	Symmetry coordinate
A_g	$\frac{1}{\sqrt{2}}(r_1 + r_5)$ $\frac{1}{\sqrt{2}}(r_2 + r_4)$ r_3 $\frac{1}{\sqrt{2}}(\theta_1 + \theta_4)$ $\frac{1}{\sqrt{2}}(\theta_2 + \theta_3)$
B_g	τ
A_u	$\frac{1}{\sqrt{2}}(\delta_1 + \delta_2)$ $\frac{1}{\sqrt{2}}(\delta_1 - \delta_2)$
B_u	$\frac{1}{\sqrt{2}}(r_1 - r_5)$ $\frac{1}{\sqrt{2}}(r_2 - r_4)$ $\frac{1}{\sqrt{2}}(\theta_1 - \theta_4)$ $\frac{1}{\sqrt{2}}(\theta_2 - \theta_3)$

5.16 Consider the internal coordinates:

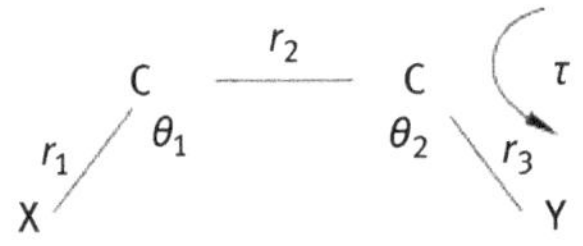

The point group is C_s. The internal coordinates and their symmetries are:

Symmetry	Coordinate
A'	$r_1, r_2, r_3, \theta_1, \theta_2$
A''	τ

The corresponding force field can be readily written down.

Chapter 7

7.2 The level energies are $\alpha+2\beta,\ \alpha+\beta\ \alpha,\ \alpha-\beta,\ \alpha-2\beta$
The level symmetries are $B_{1u},\ B_{2g},\ A_u,\ B_{3g},\ B_{1u},\ B_{2g}$
The wave functions are $1/\sqrt{12}[(\phi_1+\phi_2+\phi_3+\phi_4)+2(\phi_5+\phi_6)]$
$1/\sqrt{6}[(\phi_1+\phi_2-\phi_3-\phi_4)+(\phi_5-\phi_6)]$, $1/2(\phi_1-\phi_2+\phi_3-\phi_4)$, $1/2(\phi_1-\phi_2-\phi_3+\phi_4)$, $1/\sqrt{6}[\phi_1+\phi_2+\phi_3+\phi_4-\phi_5-\phi_6]$, $1/\sqrt{12}[(\phi_1+\phi_2-\phi_3-\phi_4)-2(\phi_5-\phi_6)]$

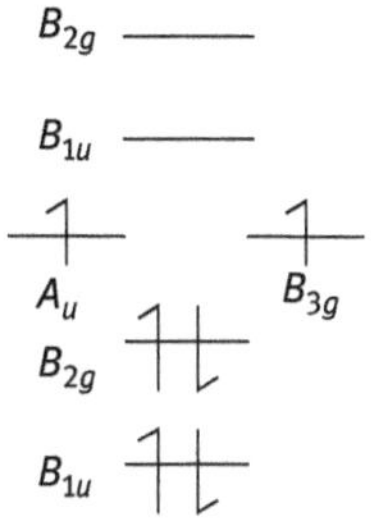

The ground state symmetry is $A_u \bullet B_{3g} = B_{3u}$. The symmetry of the excited state is $A_u \bullet B_{1u} = B_{1g}$ or $B_{3g} \bullet B_{1u} = B_{2u}$. Since $B_{3u} \bullet B_{1g} = B_{2u}$ is of y polarization, there can be transition. For $B_{3u} \bullet B_{2u} = B_{1g}$, there is no transition.

Chapter 8

8.1 $I_{Y(ZX)Y} = \frac{16\pi^4\nu^4}{c^4}\alpha_{XZ}^2 I_0$.

8.4 The possible configurations of BF_3 are pyramid C_{3v} or planar D_{3h}. The 888 cm^{-1} mode has no isotope effect, which means that B atom is stationary in this mode. This is impossible for C_{3v} configuration. Hence, the configuration of BF_3 is of point group D_{3h} [Ref.[1] of Chapter 2 vol. II, p. 298].

8.5 For Q_a, if polarizability is decided more by C=C, the relation is as below. If it is more by C–R, then it is like that of Q_b.

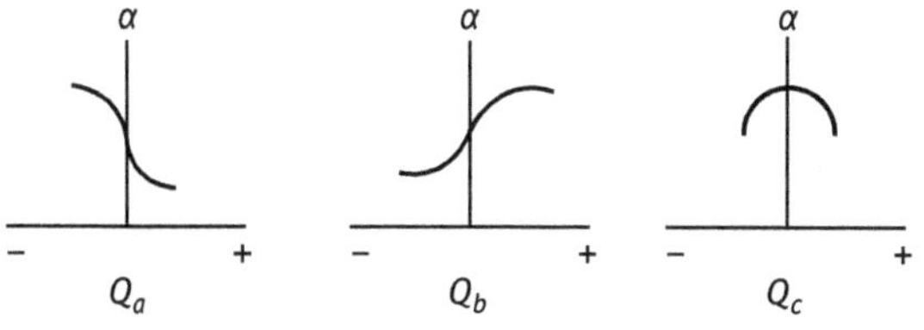

The first two modes are Raman active. The third one is i.r. active.

8.6 Raman active: $2A_{1g}, E_{1g}, 4E_{2g}$
i.r. active: $A_{2u}, 3E_{1u}$ 。

Appendix Character tables

C_s	E	σ_h		
A′	1	1	x, y, R_z	x^2, y^2 z^2, xy
A″	1	−1	z, R_x, R_y	yz, zx

C_i	E	i		
A_g	1	1	R_x, R_y, R_z	$x^2, y^2, z^2,$ xy, xz, yz
A_u	1	−1	x, y, z	

C_2	E	C_2		
A	1	1	z, R_z	x^2, y^2, z^2, xy
B	1	−1	x, y, R_x, R_y	yz, xz

C_3	E	C_3	C_3^2		$\varepsilon = \exp(2\pi i/3)$
A	1	1	1	z, R_z	x^2+y^2, z^2
E	$\left\{\begin{matrix}1\\1\end{matrix}\right.$	$\begin{matrix}\epsilon\\\epsilon^*\end{matrix}$	$\left.\begin{matrix}\epsilon^*\\\epsilon\end{matrix}\right\}$	$(x, y)\ (R_x, R_y)$	$(x^2-y^2, xy)\ (yz, xz)$

C_4	E	C_4	C_2	C_4^3		
A	1	1	1	1	z, R_z	x^2+y^2, z^2
B	1	−1	1	−1		x^2-y^2, xy
E	$\left\{\begin{matrix}1\\1\end{matrix}\right.$	$\begin{matrix}i\\-i\end{matrix}$	$\begin{matrix}-1\\-1\end{matrix}$	$\left.\begin{matrix}-i\\i\end{matrix}\right\}$	$(x, y)\ (R_x, R_y)$	(yz, xz)

C_5	E	C_5	C_5^2	C_5^3	C_5^4		$\varepsilon = \exp(2\pi i/5)$
A	1	1	1	1	1	z, R_z	x^2+y^2, z^2
E_1	$\left\{\begin{matrix}1\\1\end{matrix}\right.$	$\begin{matrix}\epsilon\\\epsilon^*\end{matrix}$	$\begin{matrix}\epsilon^2\\\epsilon^{2*}\end{matrix}$	$\begin{matrix}\epsilon^{2*}\\\epsilon^2\end{matrix}$	$\left.\begin{matrix}\epsilon^*\\\epsilon\end{matrix}\right\}$	$(x, y)\ (R_x, R_y)$	(yz, xz)
E_2	$\left\{\begin{matrix}1\\1\end{matrix}\right.$	$\begin{matrix}\epsilon^2\\\epsilon^{2*}\end{matrix}$	$\begin{matrix}\epsilon^*\\\epsilon\end{matrix}$	$\begin{matrix}\epsilon\\\epsilon^*\end{matrix}$	$\left.\begin{matrix}\epsilon^{2*}\\\epsilon^2\end{matrix}\right\}$		(x^2-y^2, xy)

https://doi.org/10.1515/9783110625097-015

C_6	E	C_6	C_3	C_2	C_3^2	C_6^5		$\varepsilon = \exp(2\pi i/6)$
A	1	1	1	1	1	1	z, R_z	x^2+y^2, z^2
B	1	−1	1	−1	1	−1		
E_1	1	ϵ	$-\epsilon^*$	−1	$-\epsilon$	ϵ^*	(x, y) (R_x, R_y)	(yz, xz)
	1	ϵ^*	$-\epsilon$	−1	$-\epsilon^*$	ϵ		
E_2	1	$-\epsilon^*$	$-\epsilon$	1	$-\epsilon^*$	$-\epsilon$		(x^2-y^2, xy)
	1	$-\epsilon$	$-\epsilon^*$	1	$-\epsilon$	$-\epsilon^*$		

C_7	E	C_7	C_7^2	C_7^3	C_7^4	C_7^5	C_7^6		$\varepsilon = \exp(2\pi i/7)$
A	1	1	1	1	1	1	1	z, R_z	x^2+y^2, z^2
E_1	1	ϵ	ϵ^2	ϵ^3	ϵ^{3*}	ϵ^{2*}	ϵ^*	(x, y) (R_x, R_y)	(yz, xz)
	1	ϵ^*	ϵ^{2*}	ϵ^{3*}	ϵ^3	ϵ^2	ϵ		
E_2	1	ϵ^2	ϵ^{3*}	ϵ^*	ϵ	ϵ^3	ϵ^{2*}		(x^2-y^2, xy)
	1	ϵ^{2*}	ϵ^3	ϵ	ϵ^*	ϵ^{3*}	ϵ^2		
E_3	1	ϵ^3	ϵ^*	ϵ^2	ϵ^{2*}	ϵ	ϵ^{3*}		
	1	ϵ^{3*}	ϵ	ϵ^{2*}	ϵ^2	ϵ^*	ϵ^3		

C_8	E	C_8	C_4	C_2	C_4^3	C_8^3	C_8^5	C_8^7		$\varepsilon = \exp(2\pi i/8)$
A	1	1	1	1	1	1	1	1	z, R_z	x^2+y^2, z^2
B	1	−1	1	1	1	−1	−1	−1		
E_1	1	ϵ	i	−1	$-i$	$-\epsilon^*$	$-\epsilon$	ϵ^*	(x, y) (R_x, R_y)	(yz, xz)
	1	ϵ^*	$-i$	−1	i	$-\epsilon$	$-\epsilon^*$	ϵ		
E_2	1	i	−1	1	−1	$-i$	i	$-i$		(x^2-y^2, xy)
	1	$-i$	−1	1	−1	i	$-i$	i		
E_3	1	$-\epsilon$	i	−1	$-i$	ϵ^*	ϵ	$-\epsilon^*$		
	1	$-\epsilon^*$	$-i$	−1	i	ϵ	ϵ^*	$-\epsilon$		

D_2	E	$C_2(z)$	$C_2(y)$	$C_2(x)$		
A	1	1	1	1		x^2, y^2, z^2
B_1	1	1	−1	−1	z, R_z	xy
B_2	1	−1	1	−1	y, R_y	xz
B_3	1	−1	−1	1	x, R_x	yz

D_3	E	$2C_3$	$3C_2$		
A_1	1	1	1		x^2+y^2, z^2
A_2	1	1	−1	z, R_z	
E	2	−1	0	(x, y) (R_x, R_y)	(x^2-y^2, xy) (xz, yz)

D_4	E	$2C_4$	$C_2(=C_4^2)$	$2C_2'$	$2C''_2$		
A_1	1	1	1	1	1		x^2+y^2, z^2
A_2	1	1	1	−1	−1	z, R_z	
B_1	1	−1	1	1	−1		x^2-y^2
B_2	1	−1	1	−1	1		xy
E	2	0	−2	0	0	(x, y) (R_x, R_y)	(xz, yz)

D_5	E	$2C_5$	$2C_5^2$	$5C_2$		
A_1	1	1	1	1		x^2+y^2, z^2
A_2	1	1	1	−1	z, R_z	
E_1	2	2cos72°	2cos144°	0	(x, y) (R_x, R_y)	(xz, yz)
E_2	2	2cos144°	2cos72°	0		(x^2-y^2, xy)

D_6	E	$2C_6$	$2C_3$	C_2	$3C_2'$	$3C''_2$		
A_1	1	1	1	1	1	1		x^2+y^2, z^2
A_2	1	1	1	1	−1	−1	z, R_z	
B_1	1	−1	1	−1	1	−1		
B_2	1	−1	1	−1	−1	1		
E_1	2	1	−1	−2	0	0	$(x,y)(R_x, R_y)$	(xz, yz)
E_2	2	−1	−1	2	0	0		(x^2-y^2, xy)

C_{2v}	E	C_2	$\sigma_v(\mathbf{xz})$	$\sigma_v(\mathbf{yz})$		
A_1	1	1	1	1	z	x^2, y^2, z^2
A_2	1	1	−1	−1	R_z	xy
B_1	1	−1	1	−1	x, R_y	xz
B_2	1	−1	−1	1	y, R_x	yz

C_{3v}	E	$2C_3$	$3\sigma_v$		
A_1	1	1	1	z	x^2+y^2, z^2
A_2	1	1	−1	R_z	
E	2	−1	0	(x, y) (R_x, R_y)	$(x^2-y^2, x\,y)$ (xz, yz)

C_{4v}	E	$2C_4$	C_2	$2\sigma_v$	$2\sigma_d$		
A_1	1	1	1	1	1	z	x^2+y^2, z^2
A_2	1	1	1	−1	−1	R_z	
B_1	1	−1	1	1	−1		x^2-y^2
B_2	1	−1	1	−1	1		xy
E	2	0	−2	0	0	(x, y) (R_x, R_y)	(xz, yz)

C_{5v}	E	$2C_5$	$2C_5^2$	$5\sigma_v$		
A_1	1	1	1	1	z	x^2+y^2, z^2
A_2	1	1	1	−1	R_z	
E_1	2	2cos72°	2cos144°	0	(x, y) (R_x, R_y)	(xz, yz)
E_2	2	2cos144°	2cos72°	0		(x^2-y^2, xy)

C_{6v}	E	$2C_6$	$2C_3$	C_2	$3\sigma_v$	$3\sigma_d$		
A_1	1	1	1	1	1	1	z	x^2+y^2, z^2
A_2	1	1	1	1	−1	−1	R_z	
B_1	1	−1	1	−1	1	−1		
B_2	1	−1	1	−1	−1	1		
E_1	2	1	−1	−2	0	0	(x, y) (R_x, R_y)	(xz, yz)
E_2	2	−1	−1	2	0	0		(x^2-y^2, xy)

C_{2h}	E	C_2	i	σ_h		
A_g	1	1	1	1	R_z	x^2, y^2, z^2, xy
B_g	1	−1	1	−1	R_x, R_y	xz, yz
A_u	1	1	−1	−1	z	
B_u	1	−1	−1	1	x, y	

C_{3h}	E	C_3	C_3^2	σ_h	S_3	S_3^5		$\epsilon = \exp(2\pi i/3)$
A'	1	1	1	1	1	1	R_z	x^2+y^2, z^2
E'	1	ϵ	ϵ^*	1	ϵ	ϵ^*	(x, y)	(x^2-y^2, xy)
	1	ϵ^*	ϵ	1	ϵ^*	ϵ		
A''	1	1	1	−1	−1	−1	z	
E''	1	ϵ	ϵ^*	−1	$-\epsilon$	$-\epsilon^*$	(R_x, R_y)	(xz, yz)
	1	ϵ^*	ϵ	−1	$-\epsilon^*$	$-\epsilon$		

C_{4h}	E	C_4	C_2	C_4^3	i	S_4^3	σ_h	S_4		
A_g	1	1	1	1	1	1	1	1	R_z	x^2+y^2, z^2
B_g	1	−1	1	−1	1	−1	1	−1		x^2-y^2, xy
E_g	$\begin{cases}1\\1\end{cases}$	i $-i$	−1 −1	$-i$ i	1 1	i $-i$	−1 −1	$\left.\begin{matrix}-i\\i\end{matrix}\right\}$	(R_x, R_y)	(xz, yz)
A_u	1	1	1	1	−1	−1	−1	−1	z	
B_u	1	−1	1	−1	−1	1	−1	1		
E_u	$\begin{cases}1\\1\end{cases}$	i $-i$	−1 −1	$-i$ i	−1 −1	$-i$ i	1 1	$\left.\begin{matrix}i\\-i\end{matrix}\right\}$	(x, y)	

C_{5h}	E	C_5	C_5^2	C_5^3	C_5^4	σ_h	S_5	S_5^7	S_5^3	S_5^9		$\varepsilon = \exp(2\pi i/5)$
A′	1	1	1	1	1	1	1	1	1	1	R_z	x^2+y^2, z^2
E'_1	1	ϵ	ϵ^2	ϵ^{2*}	ϵ^*	1	ϵ	ϵ^2	ϵ^{2*}	ϵ^*	(x, y)	
	1	ϵ^*	ϵ^{2*}	ϵ^2	ϵ	1	ϵ^*	ϵ^{2*}	ϵ^2	ϵ		
E'_2	1	ϵ^2	ϵ^*	ϵ	ϵ^{2*}	1	ϵ^2	ϵ^*	ϵ	ϵ^{2*}		(x^2-y^2, xy)
	1	ϵ^{2*}	ϵ	ϵ^*	ϵ^2	1	ϵ^{2*}	ϵ	ϵ^*	ϵ^2		
A″	1	1	1	1	1	−1	−1	−1	−1	−1	z	
E''_1	1	ϵ	ϵ^2	ϵ^{2*}	ϵ^*	−1	$-\epsilon$	$-\epsilon^2$	$-\epsilon^{2*}$	$-\epsilon^*$	(R_x, R_y)	(xz, yz)
	1	ϵ^*	ϵ^{2*}	ϵ^2	ϵ	−1	$-\epsilon^*$	$-\epsilon^{2*}$	$-\epsilon^2$	$-\epsilon$		
E''_2	1	ϵ^2	ϵ^*	ϵ	ϵ^{2*}	−1	$-\epsilon^2$	$-\epsilon^*$	$-\epsilon$	$-\epsilon^{2*}$		
	1	ϵ^{2*}	ϵ	ϵ^*	ϵ^2	−1	$-\epsilon^{2*}$	$-\epsilon$	$-\epsilon^*$	$-\epsilon^2$		

C_{6h}	E	C_6	C_3	C_2	C_3^2	C_6^5	i	S_3^5	S_6^5	σ_h	S_6	S_3		$\epsilon = \exp(2\pi i/6)$
A_g	1	1	1	1	1	1	1	1	1	1	1	1	R_z	x^2+y^2, z^2
B_g	1	−1	1	−1	1	−1	1	−1	1	−1	1	−1		
E_{1g}	1	ϵ	$-\epsilon^*$	−1	$-\epsilon$	ϵ^*	1	ϵ	$-\epsilon^*$	−1	$-\epsilon$	ϵ^*	(R_x, R_y)	(xz, yz)
	1	ϵ^*	ϵ	−1	$-\epsilon^*$	ϵ	1	ϵ^*	$-\epsilon$	−1	$-\epsilon^*$	ϵ		
E_{2g}	1	$-\epsilon^*$	$-\epsilon$	1	$-\epsilon^*$	$-\epsilon$	1	$-\epsilon^*$	$-\epsilon$	1	$-\epsilon^*$	$-\epsilon$		(x^2-y^2, xy)
	1	$-\epsilon$	$-\epsilon^*$	1	$-\epsilon$	$-\epsilon^*$	1	$-\epsilon$	$-\epsilon^*$	1	$-\epsilon$	$-\epsilon^*$		
A_u	1	1	1	1	1	1	−1	−1	−1	−1	−1	−1	z	
B_u	1	−1	1	−1	1	−1	−1	1	−1	1	−1	1		
E_{1u}	1	ϵ	$-\epsilon^*$	−1	$-\epsilon$	ϵ^*	−1	$-\epsilon$	ϵ^*	1	ϵ	$-\epsilon^*$	(x, y)	
	1	ϵ^*	$-\epsilon$	−1	$-\epsilon^*$	ϵ	−1	$-\epsilon^*$	ϵ	1	ϵ^*	$-\epsilon$		
E_{2u}	1	$-\epsilon^*$	$-\epsilon$	1	$-\epsilon^*$	$-\epsilon$	−1	ϵ^*	ϵ	−1	ϵ^*	ϵ		
	1	$-\epsilon$	$-\epsilon^*$	1	$-\epsilon$	$-\epsilon^*$	−1	ϵ	ϵ^*	−1	ϵ	ϵ^*		

D_{2h}	E	$C_2(z)$	$C_2(y)$	$C_2(x)$	i	$\sigma(xy)$	$\sigma(xz)$	$\sigma(yz)$		
A_g	1	1	1	1	1	1	1	1		x^2, y^2, z^2
B_{1g}	1	1	−1	−1	1	1	−1	−1	R_z	xy
B_{2g}	1	−1	1	−1	1	−1	1	−1	R_y	xz
B_{3g}	1	−1	−1	1	1	−1	−1	1	R_x	yz
A_u	1	1	1	1	−1	−1	−1	−1		
B_{1u}	1	1	−1	−1	−1	−1	1	1	z	
B_{2u}	1	−1	1	−1	−1	1	−1	1	y	
B_{3u}	1	−1	−1	1	−1	1	1	−1	x	

D_{3h}	E	$2C_3$	$3C_2$	σ_h	$2S_3$	$3\sigma_v$		
A_1'	1	1	1	1	1	1		$x^2 + y^2, z^2$
A_2'	1	1	−1	1	1	−1	R_z	
E'	2	−1	0	2	−1	0	(x, y)	(x^2-y^2, xy)
A''_1	1	1	1	−1	−1	−1		
A''_2	1	1	−1	−1	−1	1	z	
E''	2	−1	0	−2	1	0	(R_x, R_y)	(xz, yz)

D_{4h}	E	$2C_4$	C_2	$2C_2'$	$2C''_2$	i	$2S_4$	σ_h	$2\sigma_v$	$2\sigma_d$		
A_{1g}	1	1	1	1	1	1	1	1	1	1		$x^2 + y^2, z^2$
A_{2g}	1	1	1	−1	−1	1	1	1	−1	−1	R_z	
B_{1g}	1	−1	1	1	−1	1	−1	1	1	−1		x^2-y^2
B_{2g}	1	−1	1	−1	1	1	−1	1	−1	1		xy
E_g	2	0	−2	0	0	2	0	−2	0	0	(R_x, R_y)	(xz, yz)
A_{1u}	1	1	1	1	1	−1	−1	−1	−1	−1		
A_{2u}	1	1	1	−1	−1	−1	−1	−1	1	1	z	
B_{1u}	1	−1	1	1	−1	−1	1	−1	−1	1		
B_{2u}	1	−1	1	−1	1	−1	1	−1	1	−1		
E_u	2	0	−2	0	0	−2	0	2	0	0	(xy)	

D_{5h}	E	$2C_5$	$2C_5^2$	$5C_2$	σ_h	$2S_5$	$2S_5^3$	$5\sigma_v$		
A_1'	1	1	1	1	1	1	1	1		$x^2 + y^2, z^2$
A_2'	1	1	1	−1	1	1	1	−1	R_z	
E_1'	2	2cos72°	2cos144°	0	2	2cos72°	2cos144°	0	(x, y)	
E_2'	2	2cos144°	2cos72°	0	2	2cos144°	2cos72°	0		(x^2-y^2, xy)
A''_1	1	1	1	1	−1	−1	−1	−1		
A''_2	1	1	1	−1	−1	−1	−1	1	z	
E''_1	2	2cos72°	2cos144°	0	−2	−2cos72°	−2cos144°	0	(R_x, R_y)	(xz, yz)
E''_2	2	2cos144°	2cos72°	0	−2	−2 cos144°	−2cos72°	0		

D_{6h}	E	$2C_6$	$2C_3$	C_2	$3C'_2$	$3C''_2$	i	$2S_3$	$2S_6$	σ_h	$3\sigma_d$	$3\sigma_v$		
A_{1g}	1	1	1	1	1	1	1	1	1	1	1	1		x^2+y^2, z^2
A_{2g}	1	1	1	1	−1	−1	1	1	1	1	−1	−1	R_z	
B_{1g}	1	−1	1	−1	1	−1	1	−1	1	−1	1	−1		
B_{2g}	1	−1	1	−1	−1	1	1	−1	1	−1	−1	1		
E_{1g}	2	1	−1	−2	0	0	2	1	−1	−2	0	0	(R_x, R_y)	(xz, yz)
E_{2g}	2	−1	−1	2	0	0	2	−1	−1	2	0	0		(x^2-y^2, xy)
A_{1u}	1	1	1	1	1	1	−1	−1	−1	−1	−1	−1		
A_{2u}	1	1	1	1	−1	−1	−1	−1	−1	−1	1	1	z	
B_{1u}	1	−1	1	−1	1	−1	−1	1	−1	1	−1	1		
B_{2u}	1	−1	1	−1	−1	1	−1	1	−1	1	1	−1		
E_{1u}	2	1	−1	−2	0	0	−2	−1	1	2	0	0	(x, y)	
E_{2u}	2	−1	−1	2	0	0	−2	1	1	−2	0	0		

D_{2d}	E	$2S_4$	C_2	$2C'_2$	$2\sigma_d$		
A_1	1	1	1	1	1		x^2+y^2, z^2
A_2	1	1	1	−1	−1	R_z	
B_1	1	−1	1	1	−1		x^2-y^2
B_2	1	−1	1	−1	1	z	xy
E	2	0	−2	0	0	$(x\ y)$; (R_x, R_y)	(xz, yz)

D_{3d}	E	$2C_3$	$3C_2$	i	$2S_6$	$3\sigma_d$		
A_{1g}	1	1	1	1	1	1		x^2+y^2, z^2
A_{2g}	1	1	−1	1	1	−1	R_z	
E_g	2	−1	0	2	−1	0	(R_x, R_y)	(x^2-y^2, xy) (xz, yz)
A_{1u}	1	1	1	−1	−1	−1		
A_{2u}	1	1	−1	−1	−1	1	z	
E_u	2	−1	0	−2	1	0	(x, y)	

D_{4d}	E	$2S_8$	$2C_4$	$2S_8^3$	C_2	$4C'_2$	$4\sigma_d$		
A_1	1	1	1	1	1	1	1		x^2+y^2, z^2
A_2	1	1	1	1	1	−1	−1	R_z	
B_1	1	−1	1	−1	1	1	−1		
B_2	1	−1	1	−1	1	−1	1	z	
E_1	2	$\sqrt{2}$	0	$-\sqrt{2}$	−2	0	0	(x, y)	
E_2	2	0	−2	0	2	0	0		(x^2-y^2, xy)
E_3	2	$-\sqrt{2}$	0	$\sqrt{2}$	−2	0	0	(R_x, R_y)	(xz, yz)

D_{5d}	E	$2C_5$	$3C_5^2$	$5\,C_2$	i	$2S_{10}^3$	$2S_{10}$	$5\sigma_d$		
A_{1g}	1	1	1	1	1	1	1	1		x^2+y^2, z^2
A_{2g}	1	1	1	−1	1	1	1	−1	R_z	
E_{1g}	2	2cos72°	2cos144°	0	2	2cos72°	2cos144°	0	(R_x, R_y)	(xz, yz)
E_{2g}	2	2cos144°	2cos72°	0	2	2cos144°	2cos72°	0		(x^2-y^2, xy)
A_{1u}	1	1	1	1	−1	−1	−1	−1		
A_{2u}	1	1	1	−1	−1	−1	−1	1	z	
E_{1u}	2	2cos72°	2cos144°	0	−2	−2cos72°	−2cos144°	0	(x, y)	
E_{2u}	2	2cos144°	2cos72°	0	−2	−2cos144°	−2cos72°	0		

D_{6d}	E	$2S_{12}$	$2C_6$	$2S_4$	$2C_3$	$2S_{12}^5$	C_2	$6C_2'$	$6\sigma_d$		
A_1	1	1	1	1	1	1	1	1	1		x^2+y^2, z^2
A_2	1	1	1	1	1	1	1	−1	−1	R_z	
B_1	1	−1	1	−1	1	−1	1	1	−1		
B_2	1	−1	1	−1	1	−1	1	−1	1	z	
E_1	2	$\sqrt{3}$	1	0	−1	$-\sqrt{3}$	−2	0	0	(x, y)	
E_2	2	1	−1	−2	−1	1	2	0	0		(x^2-y^2, xy)
E_3	2	0	−2	0	2	0	−2	0	0		
E_4	2	−1	−1	2	−1	−1	2	0	0		
E_5	2	$-\sqrt{3}$	1	0	−1	$\sqrt{3}$	−2	0	0	(R_x, R_y)	(xz, yz)

S_4	E	S_4	C_2	S_4^3		
A	1	1	1	1	R_z	x^2+y^2, z^2
B	1	−1	1	−1	z	x^2-y^2, xy
E	$\left\{\begin{matrix}1\\1\end{matrix}\right.$	$\begin{matrix}i\\-i\end{matrix}$	$\begin{matrix}-1\\-1\end{matrix}$	$\left.\begin{matrix}-i\\i\end{matrix}\right\}$	$(x, y); (R_x, R_y)$	(xz, yz)

S_6	E	C_3	C_3^2	i	S_6^5	S_6		$\varepsilon = \exp(2\pi i/3)$
A_g	1	1	1	1	1	1	R_z	x^2+y^2, z^2
E_g	$\left\{\begin{matrix}1\\1\end{matrix}\right.$	$\begin{matrix}\epsilon\\\epsilon^*\end{matrix}$	$\begin{matrix}\epsilon^*\\\epsilon\end{matrix}$	$\begin{matrix}1\\1\end{matrix}$	$\begin{matrix}\epsilon\\\epsilon^*\end{matrix}$	$\left.\begin{matrix}\epsilon^*\\\epsilon\end{matrix}\right\}$	(R_x, R_y)	$(x^2-y^2, xy)\,(xz, yz)$
A_u	1	1	1	−1	−1	−1	z	
E_u	$\left\{\begin{matrix}1\\1\end{matrix}\right.$	$\begin{matrix}\epsilon\\\epsilon^*\end{matrix}$	$\begin{matrix}\epsilon^*\\\epsilon\end{matrix}$	$\begin{matrix}-1\\-1\end{matrix}$	$\begin{matrix}-\epsilon\\-\epsilon^*\end{matrix}$	$\left.\begin{matrix}-\epsilon^*\\-\epsilon\end{matrix}\right\}$	(x, y)	

S_8	E	S_8	C_4	S_8^3	C_2	S_8^5	C_4^3	S_8^7		$\varepsilon = \exp(2\pi i/8)$
A	1	1	1	1	1	1	1	1	R_z	x^2+y^2, z^2
B	1	−1	1	−1	1	−1	1	−1	z	
E_1	$\begin{cases}1\\1\end{cases}$	$\begin{matrix}\epsilon\\\epsilon^*\end{matrix}$	$\begin{matrix}i\\-i\end{matrix}$	$\begin{matrix}-\epsilon^*\\-\epsilon\end{matrix}$	$\begin{matrix}-1\\-1\end{matrix}$	$\begin{matrix}-\epsilon\\-\epsilon^*\end{matrix}$	$\begin{matrix}-i\\i\end{matrix}$	$\left.\begin{matrix}\epsilon^*\\\epsilon\end{matrix}\right\}$	(x, y) (R_x, R_y)	
E_2	$\begin{cases}1\\1\end{cases}$	$\begin{matrix}i\\-i\end{matrix}$	$\begin{matrix}-1\\-1\end{matrix}$	$\begin{matrix}-i\\i\end{matrix}$	$\begin{matrix}1\\1\end{matrix}$	$\begin{matrix}i\\-i\end{matrix}$	$\begin{matrix}-1\\-1\end{matrix}$	$\left.\begin{matrix}-i\\i\end{matrix}\right\}$		(x^2-y^2, xy)
E_3	$\begin{cases}1\\1\end{cases}$	$\begin{matrix}-\epsilon^*\\-\epsilon\end{matrix}$	$\begin{matrix}-i\\i\end{matrix}$	$\begin{matrix}\epsilon\\\epsilon^*\end{matrix}$	$\begin{matrix}-1\\-1\end{matrix}$	$\begin{matrix}\epsilon^*\\\epsilon\end{matrix}$	$\begin{matrix}i\\-i\end{matrix}$	$\left.\begin{matrix}-\epsilon\\-\epsilon^*\end{matrix}\right\}$		(xz, yz)

T_d	E	$8C_3$	$3C_2$	$6S_4$	$6\sigma_d$		
A_1	1	1	1	1	1		$x^2+y^2+z^2$
A_2	1	1	1	−1	−1		
E	2	−1	2	0	0		$(2z^2-x^2-y^2, x^2-y^2)$
T_1	3	0	−1	1	−1	(R_x, R_y, R_z)	
T_2	3	0	−1	−1	1	(x, y, z)	(xy, xz, yz)

O_h	E	$8C_3$	$6C_2$	$6C_4$	$3C_2{=}(C_4^2)$	i	$6S_4$	$8S_6$	$3\sigma_h$	$6\sigma_d$		
A_{1g}	1	1	1	1	1	1	1	1	1	1		$x^2+y^2+z^2$
A_{2g}	1	1	−1	−1	1	1	−1	1	1	−1		
E_g	2	−1	0	0	2	2	0	−1	2	0		$(2z^2-x^2-y^2, x^2-y^2)$
T_{1g}	3	0	−1	1	−1	3	1	0	−1	−1	(R_x, R_y, R_z)	
T_{2g}	3	0	1	−1	−1	3	−1	0	−1	1		(xy, xz, yz)
A_{1u}	1	1	1	1	1	−1	−1	−1	−1	−1		
A_{2u}	1	1	−1	−1	1	−1	1	−1	−1	1		
E_u	2	−1	0	0	2	−2	0	1	−2	0		
T_{1u}	3	0	−1	1	−1	−3	−1	0	1	1	(x, y, z)	
T_{2u}	3	0	1	−1	−1	−3	1	0	1	−1		

$C_{\infty v}$	E	$2C^{\Phi}_{\infty}$	...	$_\infty\sigma_v$		
$A_1 \equiv \Sigma^+$	1	1	...	1	z	x^2+y^2, z^2
$A_2 \equiv \Sigma^-$	1	1	...	−1	R_z	
$E_1 \equiv \Pi$	2	$2\cos\varphi$	...	0	$(x, y);(R_x, R_y)$	(xz, yz)
$E_2 = \Delta$	2	$2\cos 2\varphi$	...	0		(x^2-y^2, xy)
$E_3 \equiv \varphi$	2	$2\cos 3\varphi$	...	0		
...	...	...	...	...		

$D_{\infty h}$	E	$2C_{\infty}^{\varphi}$	...	$_{\infty}\sigma_v$	i	$2S_{\infty}^{\varphi}$	...	$_{\infty}C_2$		
Σ_g^+	1	1	...	1	1	1	...	1		x^2+y^2 z^2
Σ_g^-	1	1	...	−1	1	1	...	−1	R_z	
Π_g	2	$2\cos\varphi$	...	0	2	$-2\cos\varphi$	...	0	(R_x, R_y)	(xz, yz)
Δ_g	2	$2\cos2\varphi$	...	0	2	$2\cos2\varphi$	...	0		(x^2-y^2, xy)
...	...	...	...	...	...	...	...	...		
Σ_u^+	1	1	...	1	−1	−1	...	−1	z	
Σ_u^-	1	1	...	−1	−1	−1	...	1		
Π_u	2	$2\cos\varphi$	...	0	−2	$2\cos\varphi$	...	0	(x, y)	
Δ_u	2	$2\cos2\varphi$	...	0	−2	$-2\cos2\varphi$	...	0		
...	...	...	...	...	...	...	...	...		

Index

https://doi.org/10.1515/9783110625097-016

www.ingramcontent.com/pod-product-compliance
Lightning Source LLC
Chambersburg PA
CBHW081523220326
41598CB00036B/6309